Soil Hydrological Processes in Desert Regions

Soil Hydrological Processes in Desert Regions

Editors

Ying Zhao
Jianguo Zhang
Jianhua Si
Jie Xue
Zhongju Meng

MDPI • Basel • Beijing • Wuhan • Barcelona • Belgrade • Manchester • Tokyo • Cluj • Tianjin

Editors
Ying Zhao
College of Resources and Environmental Engineering
Ludong University
Yantai
China

Jianguo Zhang
College of Natural Resources and Environment
Northwest A&F University
Yangling
China

Jianhua Si
Northwest Institute of Eco-Environment and Resources
Chinese Academy of Sciences
Lan Zhou
China

Jie Xue
State Key Laboratory of Desert and Oasis Ecology
Xinjiang Institute of Ecology and Geography, Chinese Academy of Sciences
Urumqi
China

Zhongju Meng
College of Desert Control Science and Engineering
Inner Mongolia Agricultural University
Hohhot
China

Editorial Office
MDPI
St. Alban-Anlage 66
4052 Basel, Switzerland

This is a reprint of articles from the Special Issue published online in the open access journal *Water* (ISSN 2073-4441) (available at: www.mdpi.com/journal/water/special_issues/soil_hydrological).

For citation purposes, cite each article independently as indicated on the article page online and as indicated below:

LastName, A.A.; LastName, B.B.; LastName, C.C. Article Title. *Journal Name* **Year**, *Volume Number*, Page Range.

ISBN 978-3-0365-6312-1 (Hbk)
ISBN 978-3-0365-6311-4 (PDF)

Contents

Preface to "Soil Hydrological Processes in Desert Regions"

Soil hydrology is an inter-discipline of soil science and hydrology that mainly focuses on interactive pedologic and hydrologic processes and properties. The Critical Zone is the thin layer of the Earth's terrestrial surface and near-surface environment and plays a fundamental role in sustaining life and humanity. Deserts, a unique ecosystem, becomes a more critical research area in Earth's Critical Zone framework but is relatively less managed. There is such vast literature suggesting that we could tip the climate to a more humid and productive stage if we could vegetate that desert. The significance of the desert ecosystem management requires supportive and regulatory ecosystem services, ecosystem sustainability, and a feedback loop between ecological and hydrological processes. Although the benefits of reversing desertification, preventing erosion, and providing biomass have been recognized, the effects of anthropogenic revegetation on water and carbon cycles, the critical process of the terrestrial ecosystem, are still poorly understood.

This reprint highlights the current understanding of the soil hydrological processes in desert regions, as well as the plant responses to soil water. Likewise, this reprint will not present all the aspects, such as the challenges of soil hydrological process research and its opportunities in the desert regions. It only provides notable highlights to help understand the soil hydrological processes and their application in desertification control, particularly regarding the ecological engineering approach. In this context, we especially hope this reprint will enrich soil hydrology in the desert regions for advancing critical zone science in the Anthropocene.

Ying Zhao, Jianguo Zhang, Jianhua Si, Jie Xue, and Zhongju Meng
Editors

MDPI

Editorial

Special Issue: Soil Hydrological Processes in Desert Regions: Soil Water Dynamics, Driving Factors, and Practices

Ying Zhao [1,2,*], Jianguo Zhang [3], Jianhua Si [4], Jie Xue [2,*] and Zhongju Meng [5]

[1] Yantai Key Laboratory of Coastal Ecohydrological Processes and Environmental Security, College of Resources and Environmental Engineering, Ludong University, Yantai 264025, China
[2] State Key Laboratory of Desert and Oasis Ecology, Xinjiang Institute of Ecology and Geography, Chinese Academy of Sciences, Urumqi 830011, China
[3] Key Laboratory of Plant Nutrition and the Agri-Environment in Northwest China, Ministry of Agriculture, Northwest A&F University, Xianyang 712100, China
[4] Key Laboratory of Eco-Hydrology of Inland River Basin, Northwest Institute of Eco-Environment and Resources, Chinese Academy of Sciences, Lanzhou 730000, China
[5] College of Desert Control Science and Engineering, Inner Mongolia Agricultural University, Hohhot 010018, China
* Correspondence: yzhaosoils@gmail.com (Y.Z.); xuejie11@ms.xjb.ac.cn (J.X.)

Citation: Zhao, Y.; Zhang, J.; Si, J.; Xue, J.; Meng, Z. Special Issue: Soil Hydrological Processes in Desert Regions: Soil Water Dynamics, Driving Factors, and Practices. *Water* **2022**, *14*, 2635. https://doi.org/10.3390/w14172635

Received: 22 August 2022
Accepted: 25 August 2022
Published: 26 August 2022

1. Introduction

Soil hydrology is an inter-discipline of soil science and hydrology that mainly focuses on interactive pedologic and hydrologic processes and properties [1]. The Critical Zone is the thin layer of the Earth's terrestrial surface and near-surface environment and plays a fundamental role in sustaining life and human society [2,3]. Deserts, altogether a unique ecosystem, have become a more critical research area in the Earth's Critical Zone framework but one that is less managed. There is a vast amount of literature suggesting that we could tip the climate to a more humid and productive stage if we could vegetate these desert regions [4].

The significance of desert ecosystem management requires supportive and regulatory ecosystem services, ecosystem sustainability, and a feedback loop between ecological and hydrological processes [5]. Although the benefits of reversing desertification, preventing erosion, and providing biomass have been recognized, the effects of anthropogenic revegetation on soil water and carbon cycles, among many other soil hydrological processes, are still poorly understood. Over recent years, the soil structure, water retention capacity, fertility including organic carbon, and aggregation after the conversion of native desert soil into irrigated arable land have significantly changed [6]. Assessments of soil water dynamics and their driving factors are urgently needed for artificial-vegetation restoration in desert areas and the sustainability of dryland ecosystems.

This Special Issue aims to provide the highlights of the recent advances in several aspects related to the soil hydrological processes in desert regions, such as the control of land degradation and desertification, climatic and soil–water interactions, soil–plant–water–biota processes, the biogeochemical process for C and nutrient cycling, etc.

2. Summary of the Special Issue

This Special Issue publishes eleven articles, each providing valuable insights into deeply understanding the soil hydrological processes in desert regions. All the article contributions to this Special Issue are crucial for desert ecosystem management and dryland sustainability from plant responses to soil hydrological process, soil–water interaction, and practice implication perspectives. The studies covered a broad range of topics including desert-dominant species' response to soil water dynamics in extremely dry shifting desert regions; the driving factors of soil moisture in alpine deserts, desert wetlands, and desert lakes; the biogeochemical process for C and nutrient cycling in extremely dry desert regions;

the dynamic changes of soil erosion in urban desert areas; and practices for a biological protection system in shifting-sand deserts.

In the context of dominant natural species' response to soil water dynamics in extreme dry shifting desert regions, Qin et al. [7] compared the differences in the water use characteristics between two dominant species of the Badain Jaran Desert mega-dunes—*Zygophyllum xanthoxylum* and *Artemisia ordosica*—by investigating ^{2}H and ^{18}O in plant xylem and soil water and ^{13}C in plant leaves. The results confirmed that the differences in water use between the two studied species were mainly related to their root distribution characteristics. Liu et al. [8], based on the mechanistic understanding of how plant photosynthesis responds to plant drought resistance in desert regions, analyzed the daily dynamics of gas exchange parameters and their responses to photosynthetic photon flux density at three irrigation levels for two main species, *Calligonum mongolicum* and *Haloxylon ammodendron*. The results showed that *Haloxylon ammodendron* was better adapted to drought stress than *Calligonum mongolicum*. Furthermore, in determining the optimal water-saving irrigation regime, Liu et al. [9] examined the effects of irrigation regimes on the soil water dynamics and quantified the irrigation intervals and periods based on a field test of precision irrigation control in the Taklimakan Desert Highway shelterbelt. The results displayed that combining 35 mm irrigation with 10 days was beneficial to soil water storage and plant use for *Calligonum*, while combining 35 mm irrigation with 40 days was best for *Haloxylon*. In addition, Ma et al. [10], based on the issue of the impact of artificial shelterbelt constructions with saline irrigation on the soil water characteristic curve, conducted three treatments including one under the shelterbelt, on bare land in the shelterbelt, and in shifting sandy land in the hinterland of the Taklimakan Desert. The results pointed out that the influence of organic matter and salinity affected the soil water characteristic curves, which should be considered in future modeling.

Given the lacking degree of investigation on the driving factors of soil moisture in alpine deserts, desert wetlands, and desert lakes, Zhang et al. [11] examined the spatial heterogeneity and driving factors of soil gravimetric water content in the typical alpine valley desert of the Qinghai–Tibet Plateau using geostatistical analysis and the geographical detector method. The results indicated that the spatial heterogeneity and its driving factors of the soil gravimetric water content are ranked as elevation > slope > location > vegetation > aspect. Wang et al. [12], based on the issue of less attention to soil moisture and salinity in arid desert wetlands, assessed the soil moisture and salinity in the Ebinur Lake Basin. The results demonstrated that the spatial distribution of soil moisture had a higher mutation rate and stronger heterogeneity than that of soil salinity in the Ebinur Lake Basin. Jia et al. [13] investigated the hydrochemical status and evolution of lakes in the Badain Jaran Desert to understand why diverse lake water types exist under the same desert climatic conditions. The results showed that the evaporation–crystallization reactions are the controlling factors of lakes.

In the studies on the biogeochemical process for C and nutrient cycling in extremely dry desert regions, Zhang et al. [14], based on the issue of the elusive factor determining the leaf nutrients of phreatophytes in desert regions, revealed the key factors affecting the ecological stoichiometry of desert phreatophytes in the shallow groundwater of three oases in the southern rim of the Taklimakan Desert. The results showed that groundwater depth played a critical role, which was closely related to the mineralization degree of the groundwater, the topsoil C and P concentrations, and the topsoil salt content and pH. Huang et al. [15], on the issue of the little attention paid to the effects of increasing water availability on N use strategies in desert shrub species (*C. caput-medusae*), examined the changes in plant biomass, soil N status, and plant N traits, and addressed the relationships between them in 4- and 7-month-old saplings and mature shrubs after 28 months. The results displayed that increasing water availability increased the total N uptake and N resorption from old branches to satisfy the N requirement.

In an urban desert area, Zhang et al. [16] worked on the issue of the soil erosion modulus, which is closely associated with desert evolution and desertification controls.

They analyzed the spatial and temporal dynamics of soil erosion in the North and South Mountains of Lanzhou City and the soil erosion characteristics under different environmental factors and found that the soil erosion modulus is the greatest in the pedocal of the North and South Mountains and the least in the alpine soil of the mountains.

Finally, Zhao et al. [17], based on the challenge of building a highway and its biological protection system in the Taklimakan desert, comprehensively illustrated that local saline groundwater irrigation offered potential advantages and opportunities for the growth of halophytes and sandy soil development in hyper-arid desert environments.

3. Conclusions

This Special Issue highlights the current understanding of the soil hydrological processes in desert regions, as well as the plant responses to soil water. Likewise, this Special Issue will not present all the aspects, such as the challenges of soil hydrological process research and its opportunities in the desert regions. It only provides notable highlights to help understand the soil hydrological processes and their application in desertification control, particularly regarding the ecological engineering approach. In this context, we especially hope this Special Issue will enrich soil hydrology in the desert regions for advancing critical zone science in the Anthropocene.

Author Contributions: Y.Z., J.Z., J.S., J.X. and Z.M. contributed to the design, analysis, and writing of the manuscript. All authors have read and agreed to the published version of the manuscript.

Funding: This research was supported by Shandong Province College Youth Innovation Technology Support Program (2019KJF017); the Natural Science Foundation of Shandong Province, China (ZR2019JQ12); the National Talents Project (Y472241001); and the Youth Innovation Promotion Association of the Chinese Academy of Sciences (2019430).

Acknowledgments: We would like to express our sincere gratitude to the authors that contributed to this Special Issue, the professionals that reviewed the manuscript for their effort and commitment, and the editors for their contributions. We are also grateful to the Water journal staff for their help and guidance toward this successful Special Issue on "Soil Hydrological Processes in Desert Regions: Soil Water Dynamics, Driving Factors, and Practices".

Conflicts of Interest: The authors declare no conflict of interest.

References

1. Nuttle, W.K. Is ecohydrology one idea or many? *Hydrol. Sci. J.* **2002**, *47*, 805–807. [CrossRef]
2. Brantley, S.L.; Eissenstat, D.M.; Marshall, J.A.; Godsey, S.E.; Balogh-Brunstad, Z.; Karwan, D.L.; Papuga, S.A.; Roering, J.; Dawson, T.E.; Evaristo, J.; et al. Reviews and syntheses: On the roles trees play in building and plumbing the critical zone. *Biogeosciences* **2017**, *14*, 5115–5142. [CrossRef]
3. Fan, Y.; Grant, G.; Anderson, S.P. Water within, moving through, and shaping the Earth's surface: Introducing a special issue on water in the critical zone. *Hydrol. Process.* **2019**, *33*, 3146–3151. [CrossRef]
4. Zhang, Z.; Ramstein, G.; Schuster, M.; Li, C.; Contoux, C.; Yan, Q. Aridification of the Sahara desert caused by Tethys Sea shrinkage during the Late Miocene. *Nature* **2014**, *513*, 401–404. [CrossRef] [PubMed]
5. Reynolds, J.F.; Smith, D.M.S.; Lambin, E.F.; Rozema, J.; Flowers, T. Global desertification: Building a science for dryland development. *Science* **2007**, *316*, 847–851. [CrossRef] [PubMed]
6. Xue, J.; Gui, D.; Lei, J.; Sun, H.; Liu, Y. Oasification: An unable evasive process in fighting against desertification for the sustainable development of arid and semiarid regions of China. *Catena* **2019**, *179*, 197–209. [CrossRef]
7. Qin, J.; Si, J.; Jia, B.; Zhao, C.; Zhou, D.; He, X.; Wang, C.; Zhu, X. Water Use Characteristics of Two Dominant Species in the Mega-Dunes of the Badain Jaran Desert. *Water* **2022**, *14*, 53. [CrossRef]
8. Liu, J.; Zhao, Y.; Sial, T.A.; Liu, H.; Wang, Y.; Zhang, J. Photosynthetic Responses of Two Woody Halophyte Species to Saline Groundwater Irrigation in the Taklimakan Desert. *Water* **2022**, *14*, 1385. [CrossRef]
9. Liu, J.; Zhao, Y.; Zhang, J.; Hu, Q.; Xue, J. Effects of Irrigation Regimes on Soil Water Dynamics of Two Typical Woody Halophyte Species in Taklimakan Desert Highway Shelterbelt. *Water* **2022**, *14*, 1908. [CrossRef]
10. Ma, C.; Tang, L.; Chang, W.; Jaffar, M.T.; Zhang, J.; Li, X.; Chang, Q.; Fan, J. Effect of Shelterbelt Construction on Soil Water Characteristic Curves in an Extreme Arid Shifting Desert. *Water* **2022**, *14*, 1803. [CrossRef]
11. Zhang, Z.; Yin, H.; Zhao, Y.; Wang, S.; Han, J.; Yu, B.; Xue, J. Spatial Heterogeneity and Driving Factors of Soil Moisture in Alpine Desert Using the Geographical Detector Method. *Water* **2021**, *13*, 2652. [CrossRef]

12. Wang, J.; Wang, W.; Hu, Y.; Tian, S.; Liu, D. Soil Moisture and Salinity Inversion Based on New Remote Sensing Index and Neural Network at a Salina-Alkaline Wetland. *Water* **2021**, *13*, 2762. [CrossRef]
13. Jia, B.; Si, J.; Xi, H.; Qin, J. A Characterization of the Hydrochemistry and Main Controlling Factors of Lakes in the Badain Jaran Desert, China. *Water* **2021**, *13*, 2931. [CrossRef]
14. Zhang, B.; Tang, G.; Luo, H.; Yin, H.; Zhang, Z.; Xue, J.; Huang, C.; Lu, Y.; Shareef, M.; Gao, X.; et al. Topsoil Nutrients Drive Leaf Carbon and Nitrogen Concentrations of a Desert Phreatophyte in Habitats with Different Shallow Groundwater Depths. *Water* **2021**, *13*, 3093. [CrossRef]
15. Huang, C.; Zeng, F.; Zhang, B.; Xue, J.; Zhang, S. Water Supply Increases N Acquisition and N Resorption from Old Branches in the Leafless Shrub *Calligonum caput-medusae* at the Taklimakan Desert Margin. *Water* **2021**, *13*, 3288. [CrossRef]
16. Zhang, H.; Lei, J.; Wang, H.; Xu, C.; Yin, Y. Study on dynamic changes of soil erosion in the North and South Mountains of Lanzhou. *Water* **2022**, *14*, 2388. [CrossRef]
17. Zhao, Y.; Xue, J.; Wu, N.; Hill, R.L. An Artificial Oasis in a Deadly Desert: Practices and Enlightenments. *Water* **2022**, *14*, 2237. [CrossRef]

Article

Water Use Characteristics of Two Dominant Species in the Mega-Dunes of the Badain Jaran Desert

Jie Qin [1,2], Jianhua Si [1,*], Bing Jia [1,2], Chunyan Zhao [1], Dongmeng Zhou [1,2], Xiaohui He [1,2], Chunlin Wang [1,2] and Xinglin Zhu [1,2]

1 Key Laboratory of Eco-Hydrology of Inland River Basin, Northwest Institute of Eco-Environment and Resources, Chinese Academy of Sciences, Lanzhou 730000, China; qinjie18@lzb.ac.cn (J.Q.); jiab@lzb.ac.cn (B.J.); zhaochunyang@lzb.ac.cn (C.Z.); zhoudongmeng@nieer.ac.cn (D.Z.); hexiaohui@nieer.ac.cn (X.H.); wangchunlin@nieer.ac.cn (C.W.); zxinglin@yeah.net (X.Z.)

2 Northwest Institute of Eco-Environment and Resources, University of Chinese Academy of Sciences, Beijing 100049, China

* Correspondence: jianhuas@lzb.ac.cn

Abstract: The sparse natural vegetation develops special water use characteristics to adapt to inhospitable desert areas. The water use characteristics of such plants in desert areas are not yet completely understood. In this study, we compare the differences in water use characteristics between two dominant species of the Badain Jaran Desert mega-dunes—*Zygophyllum xanthoxylum* and *Artemisia ordosica*—by investigating δ^2H and δ^{18}O in plant xylem (the organization that transports water and inorganic salts in plant stems) and soil water, and δ^{13}C in plant leaves. The results indicate that *Z. xanthoxylum* absorbed 86.5% of its water from soil layers below 90 cm during growing seasons, while *A. ordosica* derived 79.90% of its water from the 0–120 cm soil layers during growing seasons. Furthermore, the long-term leaf-level water use efficiency of *A. ordosica* (123.17 ± 2.13 μmol/mol) was higher than that of *Z. xanthoxylum* (97.36 ± 1.16 μmol/mol). The differences in water use between the two studied species were mainly found to relate to their root distribution characteristics. A better understanding of the water use characteristics of plants in desert habitats can provide a theoretical basis to assist in the selection of species for artificial vegetation restoration in arid areas.

Keywords: water sources; water use efficiency; stable isotopes; Iso-source model

Citation: Qin, J.; Si, J.; Jia, B.; Zhao, C.; Zhou, D.; He, X.; Wang, C.; Zhu, X. Water Use Characteristics of Two Dominant Species in the Mega-Dunes of the Badain Jaran Desert. *Water* **2022**, *14*, 53. https://doi.org/10.3390/w14010053

Academic Editor: Carmen Teodosiu

Received: 19 November 2021
Accepted: 22 December 2021
Published: 28 December 2021

Publisher's Note: MDPI stays neutral with regard to jurisdictional claims in published maps and institutional affiliations.

1. Introduction

The Badain Jaran Desert is home to the highest mega-dunes in the world, with sparse rainfall, an arid climate, barren soil, and strong wind and sand activity. Despite the extremely arid environmental conditions, plants of the families Chenopodiaceae, Compositae, and Gramineae can be found widely distributed across the mega-dunes, as well as *Artemisia ordosica*, *Nitraria tangutorum*, *Zygophyllum xanthoxylum*, *Psammochloa villosa*, and other shrubs with tall obvious trunks and "arbor-like" growth. Among these, *Z. xanthoxylum* and *A. ordosica* represent the most dominant species and play an important role in preventing wind erosion and fixing sand. These plants are distributed 100–250 m above the lake surface; as such, it is difficult for them to use groundwater directly. Due to the extreme aridity of the desert—with an average annual rainfall of just 80 mm—the question of how these plants access and use water to survive in such an extreme environment remains. Answers to these questions are needed to clarify the water use characteristics of the plants in the Badain Jaran Desert.

Water—whether groundwater, surface water, or atmospheric precipitation—plays a highly important role in the desert ecosystem. It represents a dominant restrictive ecological factor affecting the growth of desert plants, restricting or determining the formation and development of desert vegetation [1,2]. Moreover, the distribution of desert vegetation can be analyzed to determine regional hydrological characteristics [3]. The degradation of desert vegetation is generally thought to be caused by a decline in the groundwater level, making it impossible for plants to effectively access water [4]. However, studies show that not all desert plants use groundwater alone for their maintenance [5]. In addition to groundwater, many plants use seasonal precipitation, soil water, and condensation water (such as dew and fog) to maintain their survival [5–7] However, the relative contribution of various water sources to desert plants remains inconclusive. Many previous studies used the whole-root excavation method [8], the plant physiological and ecological index identification method [9], and stable hydrogen and oxygen isotope technology [9–12] to study differences in the water sources of desert plants between species (life forms) at different times and with various spatial distributions [2,5,12]. Few studies analyze the effect of time and spatial differences together on plant water sources, which can help to better understand plant water use. The effect of such temporal and spatial changes on plant water sources are known as water use patterns [13]. In the present study, we aimed to determine the water use patterns that allow desert plants to effectively avoid or overcome water stress, and thus survive in the extremely arid mega-dune area.

Water use patterns of desert plants determine, to a certain extent, the response of the ecosystem to changes in environmental water conditions [12]. Cyclical changes in precipitation and groundwater, climate, and groundwater level in desert systems [10] lead to temporal and spatial fluctuations in the habitat conditions of desert plants, which require desert plants to adapt to changing water availability through multiple strategies such as changing their physiological characteristics and reproduction methods [14], thereby forming special water use characteristics. Water use efficiency (WUE) is another key indicator of plant water use [12,15], which can be used to characterize the ability of plants to fix organic matter under the same water consumption conditions [16]. Current mainstream research is primarily focused on instantaneous and long-term water use efficiency at the plant leaf scale [17], the latter of which can reflect the physiological conditions of plants over long-term timescales such as months and years—typically through the use of carbon stabilization isotope ratios ($\delta^{13}C$) [18]. Numerous studies show that $\delta^{13}C$ is significantly positively correlated with water use efficiency [18,19]. Therefore, carbon stable isotope technology is currently the most common method for the analysis of the long-term water use efficiency of plants.

In this study, we compare the water use characteristics of two dominant species—*Zygophyllum xanthoxylum*, a shrub species, and *Artemisia ordosica*, a semi-shrub species—in the mega-dunes of the Badain Jaran Desert using stable isotope techniques (δ^2H, $\delta^{18}O$ and $\delta^{13}C$). We investigated δ^2H and $\delta^{18}O$ in plant xylem and soil water from various layers up to 240 cm under the dry sand layer to analyze plant water sources. Additionally, we measured $\delta^{13}C$ in plant leaves to explore the interspecific differences in leaf-level WUE. The soil water content was also investigated. We hypothesized that the two species had completely different water use characteristics. The main aims of this study were: (1) to clarify the water sources of two species and determine the seasonal variation in their characteristics, (2) to analyze seasonal variations in water use efficiency (WUE) for the two species, and (3) to compare the differences in water use characteristics between the two plants. This study comprehensively analyzed the water use characteristics of different life-type plants from the perspective of time and space, which is conducive to further exploring the survival mechanism of plants in desert habitats.

2. Materials and Methods

2.1. Study Area

The Badain Jaran Desert (39°04′15″–42°12′23″ N, 99°23′18″–104°34′02″ E) is located in the west of the Alxa Plateau in Inner Mongolia. It is the second largest desert in China, with a total area of 5.22×10^4 km^2, an east-to-west length of 442 km, and a north-to-south distance of 354 km [20]. The region has an extremely arid temperate continental climate with sparse precipitation, strong evaporation, and abundant sunshine all year round. It is very hot in summer and windy in winter and spring. The average annual precipitation is 78.1 mm [21], which is mainly concentrated in the period of June to September, with the majority occurring in July and August. The air temperature ranges from a mean of −9.1 °C in winter to a mean of 25.3 °C in summer, with a mean annual temperature of 7–8 °C. The vegetation is dominated by xerophyte and super-xerophyte shrubs and semi-shrubs with low coverage. Representative plants include *Nitraria tangutorum, Zygophyllum xanthoxylum, Reaumuria soongorica, Artemisia ordosica, Agriophyllum squarrosum*, and *Psammochloa villosa* [22]. The landform is dominated by mobile sand dunes. The soil is mostly sandy soil with weak water storage capacity. The soil texture is mainly fine sand, followed by medium sand and very fine sand. The soil is alkaline, with high salinity, low organic matter and nutrient elements (Table 1). There are about 144 lakes in the Badain Jaran Desert. The depth of the lakes is relatively shallow, and the water level fluctuates less than 0.5 m. The groundwater level in this area is between 1000–1300 m [23].

Table 1. Basic soil properties of Badain Jaran Desert.

Organic Matter (g/kg)	Total Nitrogen (g/kg)	Total Carbon (g/kg)	Total Phosphorus (g/kg)	Total Salt (g/kg)	pH	Conductivity (s/m)
3.91 ± 0.46	0.71 ± 0.04	7.83 ± 0.51	0.75 ± 0.03	10.61 ± 1.87	8.15 ± 0.06	0.63 ± 0.12

2.2. Materials

The mega-dunes near Bataan Lake (102°37′1.74″ E, 39°43′19.31″ N) were selected as the test site. The test was conducted from May to August 2020, and samples were collected from 6:00 a.m. to 9:00 a.m. on each sampling date (all typical sunny days) in the middle of each month. Considering the height of the mega-dunes and the distribution of vegetation, the entire mega-dunes were divided into three regions: the upper part (180–250 m from the base of the mega-dunes), the middle part (from 90–160 m), and the lower part (from 0–70 m). Sampling was performed in 3 sampling regions on 5 mega-dunes, for a total of 15 sampling areas (Figure 1). *Zygophyllum xanthoxylum* and *Artemisia ordosica* (Figure 2), as plants typical of the mega-dune vegetation, were selected as the research objects. *Z. xanthoxylum* is a sand shrub growing in a "tree-like" shape, with fleshy leaves and an obvious main stem. The average plant is about 70 cm tall, with average main root length and diameter of 138 cm and 2.8 cm, respectively, and average lateral root length and diameter of 136 cm and 0.8 cm, respectively. *A. ordosica* is a semi-shrub, with slightly fleshy leaves, inconspicuous main stems, and many branches. The average plant is about 50 cm tall with an average root length of approximately 70 cm. In each area, three individual plants were randomly selected as representative samples for each species, and three parallel samples were collected for each individual plant. After sample collection was completed in the middle of each month, the indoor experimental analysis was carried out.

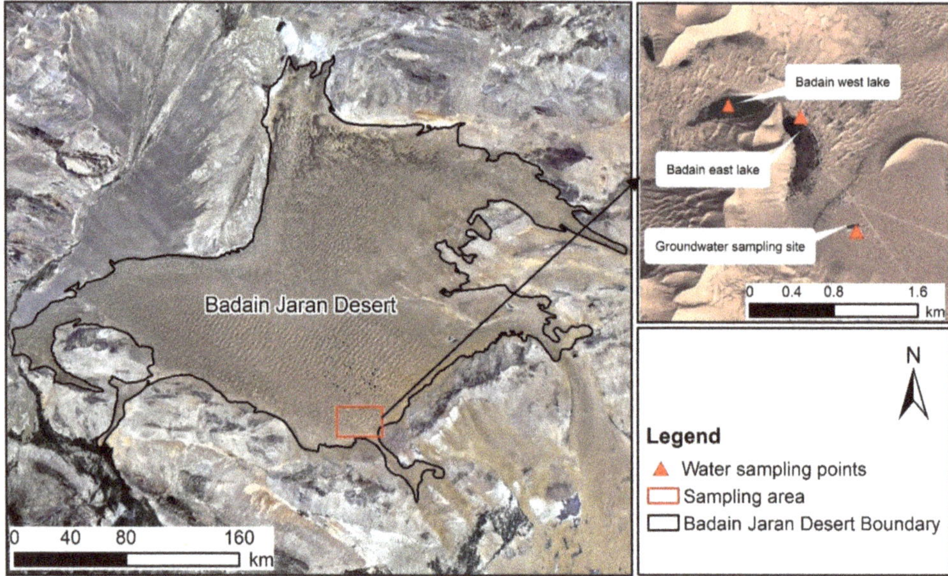

Figure 1. Location of the study area and sampling points.

Figure 2. Photographs of Zygophyllum xanthoxylum and Artemisia ordosica.

2.3. Sample Collection

Suberized and non-green twigs from different canopy directions were cut into 3–4 cm segments as samples for each species. The plant phloem tissues were also removed to avoid contamination of xylem water by isotope-enriched water [24]. Subsequently, these plant segments were immediately sealed in airtight vials, wrapped with Parafilm, stored in a portable freezer, and kept frozen at −20 °C until water extraction. Simultaneously, plant leave samples were collected in envelopes.

Soil samples were collected next to the selected plant individuals, from the emergence of the wet sand layer to 8 maximum depths (0–30, 30–60, 60–90, 90–120, 120–150, 150–180, 180–210, and 210–240 cm) by excavating soil profiles; three replicates per soil layer were performed for each species at each site. One part of each soil sample was immediately sealed in airtight vials, wrapped with Parafilm, stored in a portable freezer and kept frozen at −20 °C until water extraction for isotopic analysis. The other part was placed in a tin box to obtain the soil water content, which was determined by drying the samples at 105 °C for 24 h. After samples were collected, the sand was backfilled.

Groundwater samples were collected from the nearby wells, and lake water samples were collected from Bataan Lake (including east lake and west lake). These liquid samples were also immediately sealed in airtight vials, wrapped with Parafilm, and stored under refrigeration at −2 °C

2.4. Measurement of the δ^2H and $\delta^{18}O$ of Xylem Water and Water Sources

The cryogenic vacuum distillation system [11,25,26] was used to extract water from xylem and soil samples in the Key Laboratory of Ecohydrology of the Inland River Basin, Northwest Institute of Eco-Environment and Resources, Chinese Academy of Sciences. The whole extraction process lasted 2–3 h, depending on the sample water content [27], with an efficiency of up to 98% [5]. After extraction, the water samples were filtered with 0.22 μm pore size filters, transferred into 2 mL crimp cap vials, and stored at 4 °C until isotopic analysis. The δ^2H and δ^{18}O of all water samples were measured by an isotope ratio infrared spectroscopy (IRIS) system, namely, a Liquid Water Isotope Analyzer (LWIA, 912-0008-1001, Los Gatos Research Inc., Mountain View, CA, USA). The isotopic composition of water samples was expressed as follows:

$$\delta X(‰) = \left(R_{sample}/R_{standard} - 1\right) \times 1000 \quad (1)$$

where R_{sample} and $R_{standard}$ are the hydrogen and oxygen isotopic composition ($^2H/^1H$, $^{18}O/^{16}O$ ratios) of the sample and standard water (Vienna Standard Mean Ocean Water, V-SMOW), respectively.

2.5. Measurement of the $\delta^{13}C$ of Plant Leaves

The plant leaves were placed in an oven, deactivated at 105 °C for 25 min, dried at 70 °C for 48 h to a constant weight, then thoroughly ground and passed through a 100-mesh sieve to obtain leaf powder. Powders were sealed in containers and stored in a dry place. The δ^{13}C of plant leaves was determined by an isotope ratio mass spectrometer (IRMS) (DELTA V Advantage, Thermo Fisher Scientific, Bremen, Germany). The carbon isotopic compositions of plant leaves were expressed as in formula (1), with R_{sample} and $R_{standard}$ as the carbon isotopic composition ($^{13}C/^{12}C$ ratios) of the sample and standard (the Vienna Pee Dee Belemnite, V-PDB), respectively.

2.6. Quantification of Water Sources Used by the Plants

Since the root systems of halophytes and xerophytes undergo hydrogen isotope fractionation during the process of absorbing water [11,28], we chose to use only δ^{18}O values for the analysis and calculation of water sources. The most probable source of water absorption was found by comparing the δ^{18}O values of stem water and those of potential sources such as soil water. The Iso-source model [29] was applied to identify the proportional contributions of each water source.

2.7. Calculation of Plants Water Use Efficiency

The long-term water use efficiency of the plants was calculated according to the formulas proposed by Farquhar et al. [30]:

$$WUE = (C_a - C_i)/1.6\Delta e \quad (2)$$

$$WUE = C_a\left(1 - \frac{\delta^{13}C_a - a - \delta^{13}C_p}{b-a}\right)/1.6\Delta e \quad (3)$$

$$\delta^{13}C_a = -6.429 - 0.0060\exp[0.0217(t - 1740)] \quad (4)$$

$$\delta^{13}C_p = \delta^{13}C_a - a - (b-a)C_i/C_a \quad (5)$$

where C_a and C_i are the CO_2 pressure values in the atmosphere and leaf cells, respectively, μmol/mol; Δe is the water vapor pressure difference between the inside and outside of the leaf, MPa; $\delta^{13}C_a$ and $\delta^{13}C_p$ are the stable carbon isotope abundance values in the atmosphere and plant samples, respectively, ‰; a and b are the stable carbon isotope fractionation values during diffusion and carboxylation, respectively (approximately 4.4 ‰ and 27.0‰); and t is the year in AD, year.

2.8. Data Analysis

A one-way analysis of variance (ANOVA) was applied to detect differences in the isotopic composition and content of soil water from different soil layers. Differences in isotopic signatures, soil water content, and $\delta^{13}C$ values across different seasons, plant species, and mega-dune locations were identified using a one-way analysis of variance (ANOVA) combined with a post-hoc Tukey's least significant difference (LSD) test. The significance level of the statistical analyses was set at 0.05. Linear regression analysis was used to determine the relationship between the isotopic compositions of soil water, lake water, and groundwater, thereby analyzing the water replenishment relationship between each element. Data analyses were performed using SPSS 21.0 (SPSS Inc., Chicago, IL, USA), and figures were processed using Origin 2017 software (OriginLab Corp., Northampton, MA, USA).

3. Results

3.1. Temporal and Spatial Variations in Soil Water Content

The soil water content plots varied with soil depth and season (Figure 3). The overall soil water content in the study area is generally low, ranging from 0.41% to 2.46%, with an average value of 1.34 ± 0.53%. The soil water plots displayed no significant differences among depths but demonstrated pronounced differences among months ($p < 0.001$). Soil water content revealed a fluctuating upward trend with increase in soil depth (Figure 3). The soil water content of the 0–90 cm soil layers was highly varied—ranging from 0.50–2.40%—due to their exposure to environmental factors such as atmospheric precipitation and temperature. Therefore, the 0–90 cm soil layer can be classified as the water rapid change layer. This variation in soil water content persisted at depths of 100–180 cm, where the average water content was 0.47–2.46%. The soil water content in this layer was less affected by environmental factors but was obviously more affected by plant root water absorption, hydraulic lifting, and other activities; therefore, it can be classified as the water active layer. The soil water content of the layer between 180–240 cm had low variation and high stability, with an average water content of 1.10–2.15%. As such, it can be considered the water relatively stable layer.

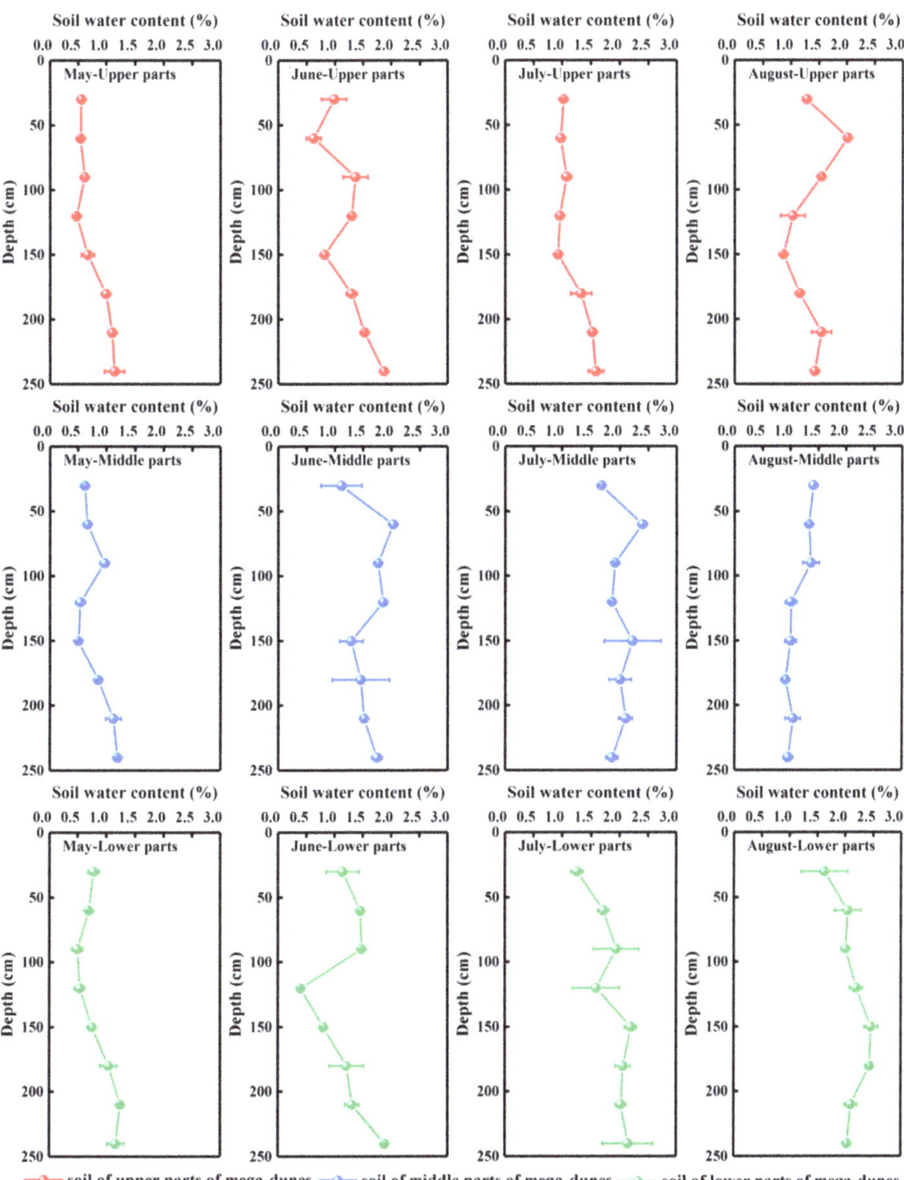

Figure 3. Vertical distribution of gravimetric soil water content of mega-dunes in the Badain Jaran Desert. Data are expressed as means ± 1SD. Rows one to three present the soil water content of the upper, middle, and lower parts of mega-dunes, respectively. Columns one to four present the soil water content in May, June, July, and August, respectively.

The soil water content changed over the months, revealing a trend of initial increase from May to July (0.80 ± 0.26%, 1.35 ± 0.41%, and 1.68 ± 1.44% for May, June, and July, respectively) followed by a decrease in August (1.54 ± 0.50%). Soil water content varied significantly with position on the mega-dunes ($p < 0.05$) with the upper parts having the lowest soil water content (1.13 ± 0.38%), the middle parts having 1.39 ± 0.51% water, and the lower parts having the highest soil water content (1.51 ± 0.60%). The soil water content of the water rapid change, water active, and water relatively stable layers also exhibited a significant difference ($p < 0.05$) among the three regions of the mega-dunes studied.

3.2. Isotopic Composition of Xylem Water and Potential Water Sources

The $\delta^{18}O$ values of groundwater and lake water varied with time (Figure 4). The $\delta^{18}O$ values of groundwater in the Badain Jaran Desert remained relatively stable across the months of the study. The $\delta^{18}O$ values of groundwater varied from −4.93‰ to −4.09‰, with an average value of −4.44 ± 0.33‰ (Figure 4). The renewal cycle of groundwater is long: it takes about 1400 years to complete a renewal of deep groundwater, while shallow groundwater also takes about 100 years. Therefore, there is little difference in the $\delta^{18}O$ values of groundwater over a short period of time.

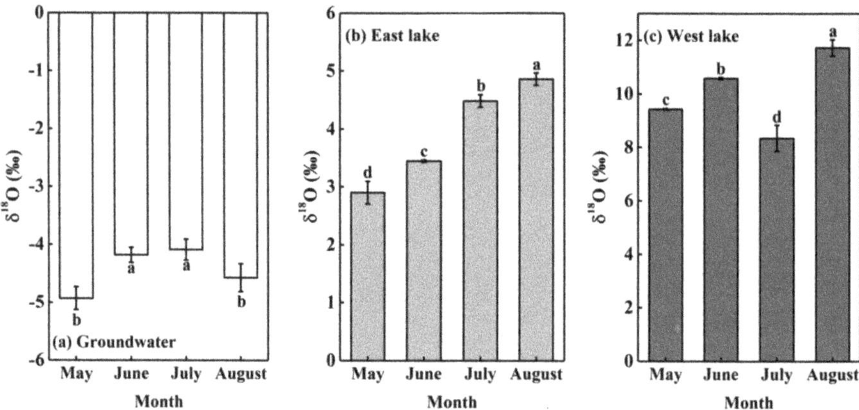

Figure 4. Seasonal variations in $\delta^{18}O$ values of groundwater and lake water. Data are expressed as means ± 1SD. Different lowercase letters express significant differences for $\delta^{18}O$ values of groundwater and lake water among months at significance level $p < 0.05$.

The $\delta^{18}O$ values of Bataan Lake water had pronounced differences across the months ($p < 0.001$) and was largest in August. The Bataan Lake is divided into east lake and west lake, and their $\delta^{18}O$ values were extremely different ($p < 0.001$). The $\delta^{18}O$ values of east lake varied from 2.90‰ to 4.86‰, with an average value of 3.93 ± 0.79‰, while the $\delta^{18}O$ values of west lake varied from 8.35‰ to 11.72‰, with an average value of 10.02 ± 1.26‰. The $\delta^{18}O$ value of west lake is much higher than that of east lake (Figure 4), mainly because east lake is a freshwater lake (total dissolved solid: 1.880 g/L) and west lake is a saltwater lake (total dissolved solid: 483.150 g/L) [31], resulting in stronger water fractionation in west lake. In addition, the δD and $\delta^{18}O$ of groundwater and lake water presented a very significant linear relationship ($\delta D = 4.491\delta^{18}O\text{-}26.462$, $R^2 = 0.994$, $p < 0.001$). The scattered points of west lake were all located on the upper right side of east lake and groundwater, while the scattered points of east lake were all located on the upper right side of groundwater (Figure 4), indicating that west lake water may be recharged by both east lake water and groundwater while east lake water was recharged by groundwater alone.

The $\delta^{18}O$ values of soil water ranged from −7.37‰ to 6.43‰, with an average value of −2.65‰. Significant differences were detected in the $\delta^{18}O$ values of soil water from different depths ($p < 0.001$) and seasons ($p < 0.001$). The $\delta^{18}O$ values of soil water gradually depleted with increase in soil depth, and the isotopic compositions of soil water in the 0–30 cm layers were enriched compared with those of the soil layers below 30 cm (Figure 5). The $\delta^{18}O$ values of soil water gradually increased from −5.02‰ in May to −4.13‰ in June, −0.93‰ in July, and finally −0.53‰ in August. This change may be due to the gradual increase in temperature from May to August and the resultant increase in soil water evaporation, resulting in the gradual enrichment of soil water $\delta^{18}O$ values. The $\delta^{18}O$ values of soil water varied significantly with position on the mega-dunes ($p < 0.05$), being lowest in the upper part (−3.24‰), −2.56‰ in the middle parts, and highest in the lower parts (−2.16‰).

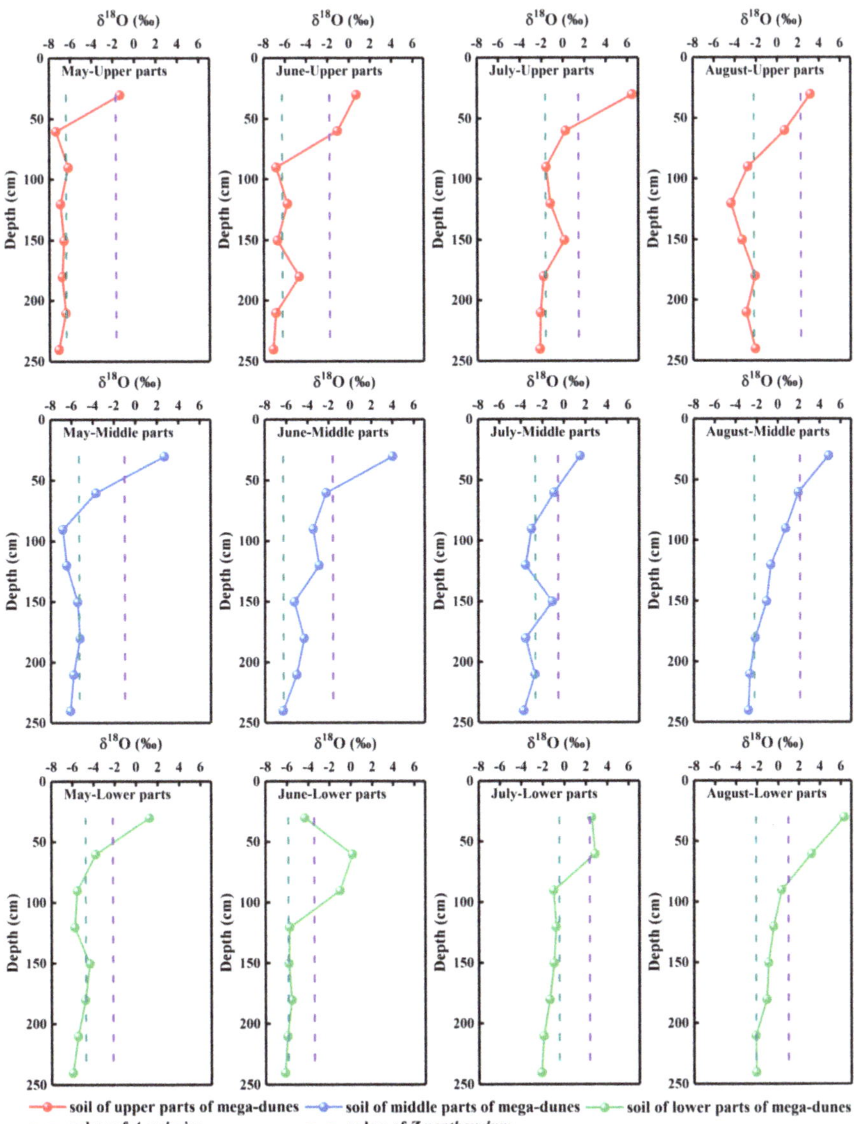

Figure 5. Spatio-temporal variations in $\delta^{18}O$ values of plant xylem water and soil water. Rows one to three present the $\delta^{18}O$ values of plant xylem water and soil water from the upper, middle, and lower parts of mega-dunes, respectively. Columns one to four present the $\delta^{18}O$ values of plant xylem water and soil water in May, June, July, and August, respectively.

The average $\delta^{18}O$ values in xylem water of *Z. xanthoxylum* ranged from −6.40‰ to −0.48‰, while those of *A. ordosica* ranged from −3.47‰ to 2.33‰. The $\delta^{18}O$ values of xylem water from the two species studied revealed pronounced variations between the different sampling dates (*Z. xanthoxylum*: $p < 0.001$; *A. ordosica*: $p < 0.01$). There was no significant difference in the xylem water $\delta^{18}O$ values for the two studied species among their different positions on the mega-dunes. However, the differences between the $\delta^{18}O$ values of xylem water from *Z. xanthoxylum* and *A. ordosica* were distinct ($p < 0.01$). The isotopic signatures in xylem water were evenly distributed and were within the range of those seen in soil water; however, they were far from those of lake water (Figure 6), indicating that plants on mega-dunes mainly absorb and use soil water and that lake water is not a potential water source. The isotopic signatures of xylem water were located at the upper right of those of groundwater (Figure 6), indicating that groundwater may contribute to plant xylem water to a certain extent; however, the plants in this study were far from groundwater level, indicating that soil water constitutes their main water source, and groundwater may replenish soil water. The intersection of the $\delta^{18}O$ value for xylem water of *Z. xanthoxylum* and that of soil water appeared within the 60–240 cm soil layer, and within the 0–60 cm soil layer for *A. ordosica* (Figure 5).

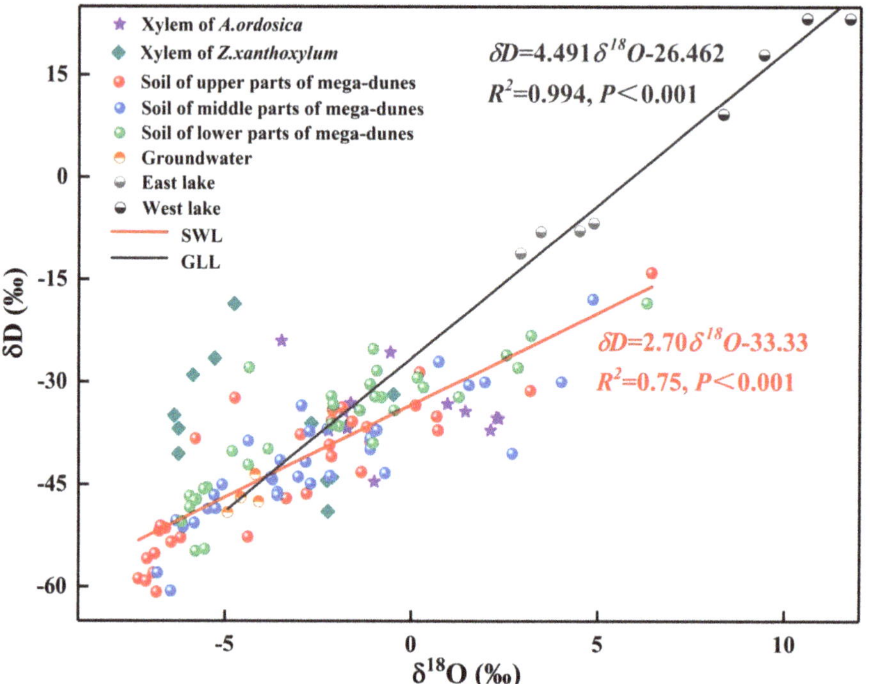

Figure 6. The δD-$\delta^{18}O$ plots of xylem water and potential water sources for *Z. xanthoxylum* and *A. ordosica*, including soil water from different soil layers, groundwater, and lake water during the sampling period. SWL is the soil water evaporation line which is fitted based on the isotopic values of soil water. GLL is fitted based on the isotopic values of groundwater, east lake water, and west lake water.

3.3. Water Use Patterns across the Growing Season

The *Z. xanthoxylum* more or less evenly absorbed water at different depths of soil in May and July. The proportional contribution of soil water from the 180–240 cm soil layers increased significantly in June and August, reaching 65.30% and 65.33%, respectively. *Z. xanthoxylum* absorbed most of its water (86.5%) from soil layers below 90 cm (Figure 7). The proportion of soil water contributed from the different soil layers was significantly different in *Z. xanthoxylum* ($p < 0.001$). *A. ordosica* derived most of its water (79.90%) from the 0–120 cm soil layers. The 0–30 cm soil layer contributed the largest percentage (64.30%) in May, and the 30–60 cm soil layer contributed the most (28.90%) in July. The proportional contributions of soil water from the 60–120 cm soil layer, from May to August, were 9.20%, 20.77%, 11.53%, and 15.17%, respectively (Figure 7). There were significant differences in the soil water contributions for *A. ordosica* from the different soil layers ($p < 0.001$). Differences in the proportional contributions from the different soil layers between *Z. xanthoxylum* and *A. ordosica* were significant, except for the 60–90 cm soil layers ($p < 0.05$). There were no distinct differences in the contribution percentages from the different soil layers for the two studied species among the different positions on the mega-dunes.

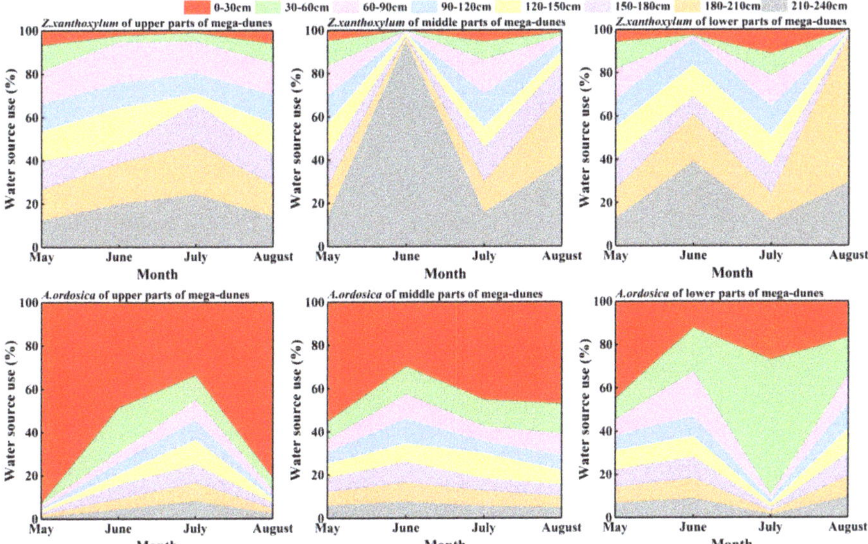

Figure 7. Seasonal variations in water uptake proportions from different soil layers for *Z. xanthoxylum* and *A. ordosica*. The data were obtained via the Iso—source model.

3.4. Temporal and Spatial Variations in Plant Water Use Efficiency

The leaf $\delta^{13}C$ values of *Z. xanthoxylum* ranged from −28.50‰ to −24.96‰, with an average value of −26.49 ± 0.10‰. *Z. xanthoxylum* had the largest leaf $\delta^{13}C$ value (−25.40 ± 0.30‰) in August, and the smallest value (−27.18 ± 0.33‰) in July (Figure 8). The leaf $\delta^{13}C$ value of *Z. xanthoxylum* had distinct differences among months ($p < 0.01$) and positions on the mega-dunes ($p < 0.001$). The leaf $\delta^{13}C$ value of *A. ordosica* ranged from −25.99‰ to −21.59‰, with an average value of −24.18 ± 0.19‰. The leaf $\delta^{13}C$ values of *A. ordosica*, from May to August, were −24.94 ± 0.82‰, −24.87 ± 0.73‰, −23.72 ± 0.19‰ and −23.20 ± 0.38‰, respectively, revealing an increasing trend over time (Figure 6). There were significant differences in the leaf $\delta^{13}C$ values of *A. ordosica* among months ($p < 0.05$) and positions on the mega-dunes ($p < 0.01$). Furthermore, the leaf $\delta^{13}C$ values of *Z. xanthoxylum* and *A. ordosica* were significantly different ($p < 0.001$).

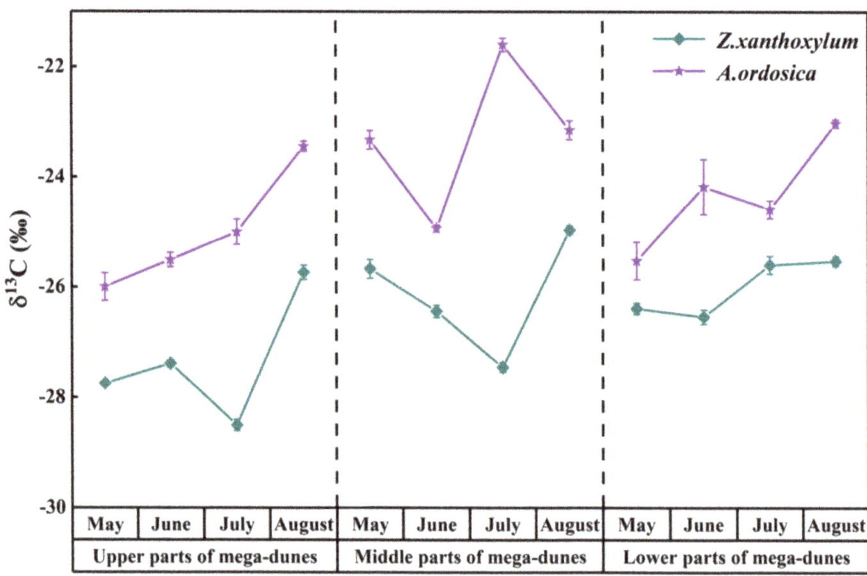

Figure 8. Spatio-temporal variations in $\delta^{13}C$ values from *Z. xanthoxylum* and *A. ordosica*. Data are expressed as means ± 1SD.

With leaf $\delta^{13}C$ values of plants, long-term water use efficiency can be obtained through the use of a formula as the same trend exists for water use efficiency and leaf $\delta^{13}C$ values. The water use efficiency of *Z. xanthoxylum* had a mean value of 97.36 ± 1.16 μmol/mol, ranging from 74.76 μmol/mol to 114.45 μmol/mol. The water use efficiency of *A. ordosica* had a mean value of 123.17 ± 2.13 μmol/mol, ranging from 102.96 μmol/mol to 152.07 μmol/mol (Figure 9). There were significant differences in the water use efficiency of both *Z. xanthoxylum* and *A. ordosica* among months ($p < 0.05$) and positions on the mega-dunes $p < 0.01$). The differences in the water use efficiency between *Z. xanthoxylum* and *A. ordosica* are pronounced ($p < 0.001$).

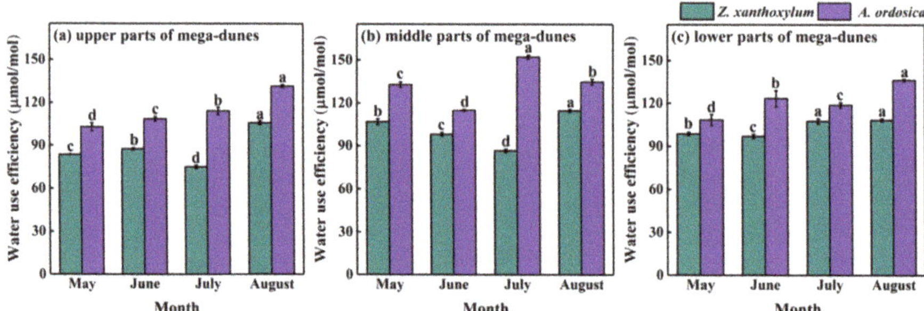

Figure 9. Spatio-temporal variations in water use efficiency for *Z. xanthoxylum* and *A. ordosica*. Data are expressed as means ± 1SD. Different lowercase letters express significant differences in plant water use efficiency among months at significance level $p < 0.05$.

4. Discussion

4.1. Variations in Soil Water Content and Soil Water Isotopic Signatures

The soil water content in this study was generally low, with an average value of less than 3.0% (Figure 3). This is mainly due to the low precipitation and the strong evaporation in this area (the evaporation is 40 to 80 times the precipitation). Otherwise, the water use patterns of desert plants determine to a certain extent the response of the ecosystem to changes in soil water conditions [12]. The long-term absorption and utilization of soil water by plants may also cause the reduction of soil water content. Strong wind-sand activity is also an important factor. The soil profiles excavated in this study were usually covered with layer of dry sand about 1 m thick, in which the soil water content was basically 0%, and water extraction could not be performed. The stability of soil water content between different soil layers was different, and the temporal change rate of soil water content in shallow and middle layers was higher than that in deep soil layers. This phenomenon is consistent with many previous studies [32,33] and is mainly related to the depth of rainfall infiltration, the intensity of soil evaporation and the depth of water absorption by plants [34].

The isotopic signatures are also mainly affected by the processes of evaporation and infiltration, mixing the original soil water and various "initial" water sources [2]. The shallow soil water is more enriched in isotopes and unstable due to more intense evaporation and unstable precipitation replenishment, while the deep soil water has weaker evaporation and is recharged by stable phreatic water at the lower interface, which lead to more negative and stable isotopic signatures. The δD of precipitation near Bataan Lake ranged from −227‰ to −20‰, while the $\delta^{18}O$ ranged from −28.5‰ to −1.3‰ [35]. Compared with the soil water isotope in this study, it was found that precipitation has a certain degree of replenishment to shallow soil water. According to the analysis in Figure 4, it was founded that groundwater (phreatic water) may recharge the soil water, indicating that there was strong water movement in the sand layers of mega-dunes.

4.2. Differences in Water Use Patterns

The distribution characteristics of roots—the water-absorbing end of plants—are closely related to the sources or the soil layers from which water is absorbed [12,36,37]. Many previous excavation studies on desert plants have demonstrated that the roots of perennials continuously appear throughout all soil layers that may be recharged once a year, and a few plants actually penetrate the groundwater table [8,10,38], but the most active water-absorbing fine roots were mainly found in deep soils [39]. The water absorption layers of the two studied species typical of the mega-dunes of the Badain Jaran Desert are also mainly determined by the distribution of their roots. *Z. xanthoxylum* has a deep root system that can extend to about 2 m below the dry sand layer so that it can evenly use soil moisture from depths of 0–240 cm. This finding is different to the results of Zhang et al. [40] and Chen et al. [41] regarding the water absorption layer (20–60 cm) of *Z. xanthoxylum* in the West Ordos Desert, which may be related to the adaption of plant roots to different habitats: the vertical distribution depth of the root system of *Z. xanthoxylum* in the mega-dunes of the Badain Jaran Desert—where the habitat is more arid—is deeper than in the West Ordos Desert. Therefore, it can be concluded that differences in habitat affect the water sources of *Z. xanthoxylum*. Plants can flexibly adjust their root distribution and other morphological characteristics according to habitat conditions and present the best phenotypes in various habitats [42]. However, the two abovementioned studies on the water use strategy of *Z. xanthoxylum* proposed that it simultaneously uses shallow and deep soil water, which is basically consistent with the conclusions of our study. The root system of *A. ordosica* is mainly distributed around 0–100 cm, so it mainly absorbs and utilizes soil water from the 0–60 cm layer, indicating that it is not the entire root system of plants that absorb water, but the active root system—distributed in the layer containing effective water—which can continuously absorb water from soil for plant growth and use [43]. Huang et al. [44] analyzed the water source of *A. ordosica* in the Tengger Desert and found that it mainly utilizes water in the upper layer (10–100 cm) of soil, and mainly uses shallow soil water in the dry season. Fu et al. [45] studied the water source of *A. ordosica* in the Mu Us sandy land and found that—while it can use soil moisture from the 0–140 cm soil layer—the water uptake proportion of the 0–60 cm soil layer was distinctly higher than that of the 60–140 cm layer. The results of these quantitative analyses on the of the water source of *A. ordosica* are basically consistent with our study. Therefore, it can be concluded that the root genetic characteristics of *A. ordosica* primarily determine its water sources in similar habitats.

Many scholars believe that plants in arid areas can flexibly use various water sources due to the dimorphism of their roots and the obvious seasonal changes in water sources of such plants [46–49]. Plants that continue to draw water from shallow soils may struggle to survive extreme drought conditions [12]. The water sources of the two plants in our study did not exhibit a uniform pattern of variation across the months they were studied. However, the two plants preferred to use more soil water from a relatively deep level, in terms of their own water range, in the middle of the growing season when their water demand was large, or the water condition was poor: *Z. xanthoxylum* significantly increased its use of soil water from the 180–240 cm soil layer, while *A. ordosica* used more water from the 30–90 cm soil layer. This may be because the Badain Jaran Desert experienced long-term drought in the winter and spring of this year, and precipitation occurred in May; however, these precipitation events were relatively small (less than 3 mm in a single event) and could not penetrate into the middle or deep soil [50]. Therefore, only the shallow or superficial moisture of the sand layer was replenished, and the plant roots distributed in this layer were stimulated to quickly absorb water. The temperature gradually increased in June and July while the precipitation was less than that of May, which made the shallow soil moisture evaporate quickly and prevented it from being recharged effectively (the precipitation reached the sand layer and quickly evaporated within a few hours). The water demand for plant growth at this time gradually increases, so plants could only switch to using more of the stable moisture from the middle and deep soil layers. This is a unique water utilization strategy for xerophyte survival in extreme environments.

Due to the dry climate of the Badain Jaran Desert, condensation, in addition to soil water, may play an important role in the growth and development of plants. Gong et al. [51] found that many plant leaves in arid areas can directly absorb condensation water and increase leaf water potential, thereby reducing the water deficit of plants. Hill et al. [52] analyzed the water sources of three xerophytes in the Negev Desert in Israel and found that the condensed water absorbed and used by plants accounted for more than 50% of their total water. In this study, only a small number of condensed water samples were collected, which led to the failure to analyze the relationship between condensed water and plant water. In future studies, the method of collecting condensed water samples should be optimized, and as many condensed water samples as possible should be collected to accurately and quantitatively analyze the contribution of condensed water to plant water.

4.3. Differences in Water Use Efficiency

Plant water use efficiency is an important index that couples leaf photosynthesis and water physiological processes, and can effectively evaluate the suitability of plant growth [15,16]. The normal physiological activity and growth of plants required high water use efficiency to be guaranteed in an extremely water-deficient desert ecosystem [12,30]. Previous studies have shown that plants in a water limited environment have higher water use efficiency that plants in an environment with sufficient water supply [14,53]. The average $\delta^{13}C$ values of *Z. xanthoxylum* and *A. ordosica* in our study were larger than those of global plants (C_3 plants), which is consistent with the conclusion that plants have relatively higher water efficiency in arid regions [53]. The water use efficiency of *A. ordosica* is higher than *Z. xanthoxylum* (Figure 7), which is mainly related to the genetic characteristics of the plant itself. In addition, the root distribution characteristics of the two plants are different, and the soil layers that they absorb water are different, resulting in differences in the degree of difficulty in obtaining water. Plants with higher water use efficiency are more advantageous in water-deficient habitats. The higher water use efficiency of *A. ordosica* also effectively explains its widespread distribution in the Badain Jaran Desert.

Plant water use efficiency is affected by many factors, including plant species, growth stage, plant internal factors (photosynthetic organ structure, leaf water potential, stomata, etc.) and external environmental factors (light, moisture, carbon dioxide concentration, air temperature, leaf temperature, etc.). In addition to the differences in species and soil water content, the spatial variations of plant water use efficiency in mega-dunes may also be related to the soil texture of mega-dunes, which has a strong correlation with the degree of water availability. In a water-restricted environment, in addition to flexible use of water sources and higher water use efficiency, plants adjusted their water use characteristics through morphological adaptation, stomatal regulation mechanism, osmotic regulation mechanism, antioxidant defense regulation mechanism, and hormone regulation mechanism.

4.4. Interaction between Different Plants in the Community

Plants that continue to use shallow or deep soil water for a long time may cause soil desiccation and degradation of ecosystem services [12,54]. Different plants that grow in the same habitat will usually absorb and utilize soil moisture from different depths [55,56], which is also reflected in our study. *Z. xanthoxylum* absorbed most of its water (86.5%) from below 90 cm soil layers, while *A. ordosica* derived most of its water (79.90%) from 0~120 soil layers. As a result, the limited soil moisture can be utilized to the maximum to meet the water demand of the community. In addition, there are perennial herbaceous plants in this habitat whose water absorption layer is about 0–60 cm, which overlaps with the water use layer of *A. ordosica*, indicating that there is a certain water competition relationship between them. Additionally, although the root system of *A. ordosica* is relatively shallower than that of *Z. xanthoxylum*, the analysis of water sources revealed that *A. ordosica* also absorbs a small or very small amount of soil water below 150 cm that its root system cannot reach. This phenomenon may be related to the water-lifting effect of *Z. xanthoxylum*'s root system [57]: the deep soil water absorbed by *Z. xanthoxylum*'s root system is released in relatively shallow soil and used by *A. ordosica*. This indicates that a variety of plants may have both a competitive and a mutually beneficial symbiotic relationship in extremely arid habitats.

The plants water use characteristics are an important basis for species selection in desert vegetation restoration or reconstruction. The amount of water sources should be considered. If the utilization of soil moisture and groundwater is not enough to maintain the normal growth of communities in this area, this may lead to a large area of vegetation degradation [58]. When selecting species, it is necessary to comprehensively consider the amount of available water resources in the area and the water use characteristics of various plants to determine the most suitable community type.

5. Conclusions

To better understand the water use characteristics of the two dominate species of the mega-dunes of the Badain Jaran Desert—*Z. xanthoxylum* and *A. ordosica*—we measured the δ^2H and δ^{18}O of plant xylem and soil water from various layers up to 240 cm under the dry sand layer and, in combination with the Iso-source model, analyzed plant water sources and the δ^{13}C of plant leaves to explore the interspecific differences in leaf-level WUE. The Iso-source model predicted that *Z. xanthoxylum* derived most of its water (86.5%) from soil layers below 90 cm, while *A. ordosica* derived most of its water (79.90%) from the 0–120 cm soil layers during growing seasons. The leaf δ^{13}C implied that the long-term leaf-level water use efficiency of *A. ordosica* is higher than that of *Z. xanthoxylum* during growing seasons. These results suggest that *A. ordosica* is more suitable for survival in this habitat. This indicates that *A. ordosica* can be selected as a sand-fixing plant for vegetation restoration and construction in sandy soil habitats with sparse rainfall and deep groundwater levels. Moreover, there existed both a competitive and a mutually beneficial symbiotic relationship between a variety of plants from the same arid habitat. This study reveals the water use mechanism of plants on mega-dunes far from the groundwater level and provides a theoretical basis to assist in the selection of species for artificial vegetation restoration in arid areas. In addition, condensation water plays a potential role in the growth and development of plants, which will be the subject of research in the future.

Author Contributions: J.Q.: Conceptualization, Data curation, Investigation, Methodology, Formal analysis, Roles/Writing—original draft, Writing—review & editing; J.S.: Conceptualization, Data curation, Formal analysis, Funding acquisition, Methodology; B.J.: Investigation; C.Z.: Conceptualization, Data curation, Formal analysis, Investigation; D.Z.: Investigation; X.H.: Investigation; C.W.: Investigation; X.Z.: Investigation. All authors have read and agreed to the published version of the manuscript.

Funding: This study was supported by the Major Science and Technology Project in Inner Mongolia Autonomous region of China (No. Zdzx2018057), the Innovation Cross Team Project of Chinese Academy of Sciences, CAS (No. JCTD-2019-19), Transformation Projects of Scientific and Technological Achievements in Inner Mongolia Autonomous region of China (No. 2021CG0046), and the National Natural Science Foundation of China (No. 42001038).

Institutional Review Board Statement: Not applicable.

Informed Consent Statement: Not applicable.

Data Availability Statement: The data that support the findings of this study are available from the corresponding author on reasonal request.

Acknowledgments: We greatly appreciate suggestions from anonymous referees for the improvement of our paper. Thanks also to the editorial staff.

Conflicts of Interest: The authors declare that they have no known competing financial interests or personal relationships that could have appeared to influence the work reported in this paper.

References

1. Fay, P.A.; Carlisle, J.D.; Knapp, A.K.; Blair, J.M.; Collins, S.L. Productivity responses to altered rainfall patterns in a C_4-dominated grassland. *Oecologia* **2003**, *137*, 245–251. [CrossRef]
2. Zhao, L.J.; Wang, X.G.; Zhang, Y.C.; Xie, C.; Liu, Q.Y.; Meng, F. Plant water use strategies in the Shapotou artificial sand-fixed vegetation of the southeastern margin of the Tengger Desert, northwestern China. *J. Mt. Sci.* **2019**, *16*, 898–908. [CrossRef]
3. Liu, Y.; Fu, B.J.; Lü, Y.H.; Gao, G.Y.; Wang, S.; Zhou, J. Linking vegetation cover patterns to hydrological responses using two process-based pattern indices at the plot scale. *Sci. China-Earth Sci.* **2013**, *56*, 1888–1898. [CrossRef]
4. Hao, X.M.; Chen, Y.N.; Li, W.H.; Guo, B.; Zhao, R.F. Hydraulic lift in *Populus euphratica Oliv.* from the desert riparian vegetation of the Tarim River Basin. *J. Arid. Environ.* **2010**, *74*, 905–911. [CrossRef]
5. Wu, X.; Zheng, X.J.; Yin, X.W.; Yue, Y.M.; Liu, R.; Xu, G.Q.; Li, Y. Seasonal variation in the groundwater dependency of two dominant woody species in a desert region of Central Asia. *Plant Soil* **2019**, *444*, 39–55. [CrossRef]
6. Kolb, T.; Hart, S.; Amundson, R. Boxelder water source and physiology at perennial and ephemeral stream sites in Arizona. *Tree Physiol.* **1997**, *17*, 151–160. [CrossRef]
7. Jacobs, A.F.G.; Heusinkveld, B.G.; Berkowicz, S.M. A simple model for potential dewfall in an arid region. *Atmos. Res.* **2002**, *64*, 285–295. [CrossRef]
8. Cannon, W.A. *The Root Habits of Desert Plants*; Carnegie Inst of Washington Publ: Washington, DC, USA, 1911; p. 96.
9. Nie, Y.P.; Chen, H.S.; Wang, K.L. Methods for determining plant water source in thin soil region: A review. *Chin. J. Appl. Ecol.* **2010**, *21*, 2427–2433.
10. Ehleringer, J.R.; Phillips, S.L.; Schuster, W.S.F.; Sandquist, D.R. Differential utilization of summer rains by desert plants. *Oecologia* **1991**, *88*, 430–434. [CrossRef]
11. Zhao, L.J.; Wang, L.X.; Cernusak, L.A.; Liu, X.H.; Xiao, H.L.; Zhou, M.X.; Zhang, S.Q. Significant difference in hydrogen isotope composition between xylem and tissue water in *Populus euphratica*. *Plant Cell Environ.* **2016**, *39*, 1848–1857. [CrossRef] [PubMed]
12. Wang, J.; Fu, B.J.; Lu, N.; Wang, S.; Zhang, L. Water use characteristics of native and exotic shrub species in the semi-arid Loess Plateau using an isotope technique. *Agric. Ecosyst. Environ.* **2019**, *276*, 55–63. [CrossRef]
13. Wang, J.; Fu, B.J.; Lu, N.; Zhang, L. Seasonal variation in water uptake patterns of three plant species based on stable isotopes in the semi-arid Loess Plateau. *Sci. Total Environ.* **2017**, *609*, 27–37. [CrossRef] [PubMed]
14. Gries, D.; Zeng, F.; Foetzki, A.; Arndt, S.K.; Bruelheide, H.; Thomas, F.M.; Zhang, X.; Runge, M. Growth and water relations of *Tamarix ramosissima* and *Populus euphratica* on Taklamakan desert dunes in relation to depth to a permanent water table. *Plant Cell Environ.* **2003**, *26*, 725–736. [CrossRef]
15. Gao, Y.; Markkanen, T.; Aurela, M.; Mammarella, I.; Thum, T.; Tsuruta, A.; Yang, H.; Aalto, T. Response of water use efficiency to summer drought in a boreal Scots pine forest in Finland. *Biogeosciences* **2017**, *14*, 4409–4422. [CrossRef]

16. Lavergne, A.; Graven, H.; De Kauwe, M.G.; Keenan, T.F.; Medlyn, B.E.; Prentice, I.C. Observed and modelled historical trends in the water-use efficiency of plants and ecosystems. *Glob. Change Biol.* **2019**, *25*, 2242–2257. [CrossRef] [PubMed]
17. Medrano, H.; Tomás, M.; Martorell, S.; Flexas, J.; Hernández, E.; Rosselló, J.; Pou, A.; Escalona, J.M.; Bota, J. From leaf to whole-plant water use efficiency (WUE) in complex canopies: Limitations of leaf WUE as a selection target. *Crop J.* **2015**, *3*, 220–228. [CrossRef]
18. Ehleringer, J.R.; Cooper, T.A. Correlations between carbon isotope ratio and microhabitat in desert plants. *Oecologia* **1988**, *76*, 562–566. [CrossRef]
19. Ebdon, J.S.; Petrovic, A.M.; Dawson, T.E. Relationship between carbon isotope discrimination, water use efficiency, and evapotranspiration in Kentucky Bluegrass. *Crop Sci.* **1998**, *38*, 157–162. [CrossRef]
20. Zhu, J.F.; Wang, N.A.; Chen, H.B.; Dong, C.Y.; Zhang, H.A. Study on the boundary and the area of Badain Jaran Desert based on remote sensing imagery. *Prog. Geogr.* **2010**, *29*, 1087–1094.
21. Ma, N.; Wang, N.A.; Zhu, J.F.; Chen, X.L.; Chen, H.B.; Dong, C.Y. Climate change around the Badain Jaran Desert in recent 50 years. *J. Desert Res.* **2011**, *31*, 1541–1547.
22. Qin, J.; Si, J.H.; Jia, B.; Zhao, C.Y.; Li, D.; Luo, H.; Ren, L.X. Study on the relationship between vegetation community characteristics and soil moisture in Badain Jaran Desert. *Arid. Zone Res.* **2021**, *38*, 207–222.
23. Zhang, J.; Wang, X.S.; Hu, X.N.; Lu, H.T.; Gong, Y.P.; Wan, L. The macro-characteristics of Groundwater flow in the Badain Jaran Desert. *J. Desert Res.* **2015**, *35*, 774–782.
24. Querejeta, J.I.; Estrada-Medina, H.; Allen, M.F.; Jiménez-Osornio, J.J. Water source partitioning among trees growing on shallow karst soils in a seasonally dry tropical climate. *Oecologia* **2007**, *152*, 26–36. [CrossRef]
25. Ehleringer, J.R.; Roden, J.; Dawson, T.E. Assessing Ecosystem-Level Water Relations through Stable Isotope Ratio Analyses. In *Methods in Ecosystem Science*; Springer: New York, NY, USA, 2000; pp. 181–198.
26. West, A.G.; Patrickson, S.J.; Ehleringer, J.R. Water extraction times for plant and soil materials used in stable isotope analysis. *Rapid Commun. Mass Spectrom.* **2006**, *20*, 1317–1321. [CrossRef]
27. West, J.B.; Bowen, G.J.; Cerling, T.E.; Ehleringer, J.R. Stable isotopes as one of nature's ecological recorders. *Trends Ecol. Evol.* **2006**, *21*, 408–414. [CrossRef] [PubMed]
28. Ellsworth, P.Z.; Williams, D.G. Hydrogen isotope fractionation during water uptake by woody xerophytes. *Plant Soil* **2007**, *291*, 93–107. [CrossRef]
29. Phillips, D.L.; Gregg, J.W. Source partitioning using stable isotopes: Coping with too many sources. *Oecologia* **2003**, *136*, 261–269. [CrossRef] [PubMed]
30. Farquhar, G.D.; O'Leary, M.H.; Berry, J.A. On the relationship between carbon isotope discrimination and the intercellular carbon dioxide concentration in leaves. *Aust. J. Plant Physiol.* **1982**, *9*, 281–292. [CrossRef]
31. Chen, L.; Wang, N.A.; Wang, H.; Dong, C.Y.; Lu, Y.; Lu, J.W. Spatial patterns of chemical parameters of lakes and groundwater in Badain Jaran Desert. *J. Desert Res.* **2012**, *32*, 531–538.
32. Penna, D.; Brocca, L.; Borga, M.; Dalla, F.G. Soil moisture temporal stability at different depths on two alpine hillslopes during wet and dry periods. *J. Hydrol.* **2013**, *477*, 55–71. [CrossRef]
33. Yang, L.; Wei, W.; Chen, L.D.; Chen, W.L.; Wang, J.L. Response of temporal variation of soil moisture to vegetation restoration in semi-arid Loess Plateau, China. *Catena* **2014**, *115*, 123–133. [CrossRef]
34. Seneviratne, S.I.; Corti, T.; Davin, E.L.; Hirschi, M.; Jaeger, E.B.; Lehner, I.; Orlowsky, B.; Teuling, A.J. Investigating soil moisture–climate interactions in a changing climate: A review. *Earth-Sci. Rev.* **2010**, *99*, 125–161. [CrossRef]
35. Cao, L.; Shen, J.M.; Nie, Z.L.; Meng, L.Q.; Liu, M.; Wang, Z. Stable isotopic characteristics of precipitation and moisture recycling in Badain Jaran Desert. *Earth Sci.* **2021**, *46*, 2973–2983.
36. Dawson, T.E.; Mambelli, S.; Plamboeck, A.H.; Templer, P.H.; Tu, K.P. Stable isotopes in plant ecology. *Annu. Rev. Ecol. Syst.* **2002**, *33*, 507–559. [CrossRef]
37. Schenk, H.J. Soil depth, plant rooting strategies and species' niches. *New Phytol.* **2008**, *178*, 223–225. [CrossRef] [PubMed]
38. Manning, S.J.; Barbour, M.G. Root systems, spatial patterns, and competition for soil moisture between two desert subshrubs. *Am. J. Bot.* **1988**, *75*, 885–893. [CrossRef]
39. Dawson, T.E.; Ehleringer, J.R. Streamside trees that do not use stream water. *Nature* **1991**, *350*, 335–337. [CrossRef]
40. Zhang, Y.M.; Gao, R.H.; Jin, H. Studies on the ecological characteristics of four bushes roots in desert of West Erdos. *J. Inn. Mong. Agric. Univ.* **2005**, *26*, 39–43.
41. Chen, J.; Xu, Q.; Gao, D.Q.; Ma, Y.B. Water use of *Helianthemum songaricum* and co-occurring plant species *Sarcozygium xanthoxylum* in Western Ordos. *Sci. Silvae Sin.* **2016**, *52*, 47–56.
42. Zhou, H.; Zhao, W.Z.; He, Z.B. Water sources of *Nitraria sibirica* and response to precipitation in two desert habitats. *Chin. J. Appl. Ecol.* **2017**, *28*, 2083–2092.
43. Donovan, L.A.; Ehleringer, J.R. Water stress and use of summer precipitation in a Great Basin shrub community. *Funct. Ecol.* **1994**, *8*, 289–297. [CrossRef]
44. Huang, L.; Zhang, Z.S.; Li, X.R. Sap flow of *Artemisia ordosica* and the influence of environmental factors in a revegetated desert area: Tengger Desert, China. *Hydrol. Processes* **2010**, *24*, 1248–1253.
45. Fu, X.; Wu, Y.S.; Zhang, Y.J. Analysis of water sources utilized by four shrubs during the growing season in the southern margin of Mu Us Sandy Land. *J. Inn. Mong. Norm. Univ.* **2020**, *49*, 245–250.

46. Dawson, T.E.; Pate, J.S. Seasonal water uptake and movement in root systems of Australian phraeatophytic plants of dimorphic root morphology: A stable isotope investigation. *Oecologia* **1996**, *107*, 13–20. [CrossRef] [PubMed]
47. Xi, B.Y.; Wang, Y.; Jia, L.M.; Bloomberg, M.; Li, G.D.; Di, N. Characteristics of fine root system and water uptake in a triploid *Populus tomentosa* plantation in the North China Plain: Implications for irrigation water management. *Agric. Water Manag.* **2013**, *117*, 83–92. [CrossRef]
48. Xing, X.; Chen, H.; Zhu, J.J.; Chen, T.T. Water sources of five dominant desert plant species in Nuomuhong area of Qaidam Basin. *Acta Ecol. Sin.* **2014**, *34*, 6277–6286.
49. Wang, J.; Zhang, R.; Yan, Y.T.; Dong, X.Q.; Li, J.M. Locating hazardous gas leaks in the atmosphere via modified genetic, MCMC and particle swarm optimization algorithms. *Atmos. Environ.* **2017**, *157*, 27–37. [CrossRef]
50. Feng, Q.; Cheng, G.D. Moisture distribution and movement in sandy lands of China. *Acta Pedol. Sin.* **1999**, 225–236.
51. Gong, X.W.; Lü, G.H.; Ran, Q.Y.; Yang, X.D. Response of vegetative growth and biomass allocation of *Lappula semiglabra* seedlings to dew gradient. *Chin. J. Appl. Ecol.* **2016**, *27*, 2257–2263.
52. Hill, A.J.; Dawson, T.E.; Shelef, O.; Rachmilevitch, S. The role of dew in Negev Desert plants. *Oecologia* **2015**, *178*, 317–327. [CrossRef] [PubMed]
53. Wright, G.C.; Rao, R.C.; Farquhar, G.D. Water-use efficiency and carbon isotope discrimination in peanut under water deficit conditions. *Crop Sci.* **1994**, *34*, 92–97. [CrossRef]
54. Chen, H.S.; Shao, M.G.; Li, Y.Y. Soil desiccation in the Loess Plateau of China. *Geoderma* **2008**, *143*, 91–100. [CrossRef]
55. Noymeir, I. Desert ecosystems: Environment and producers. *Annu. Rev. Ecol. Syst.* **1973**, *4*, 25–51. [CrossRef]
56. Darrouzet-Nardi, A.; D' Antonio, C.M.; Dawson, T.E. Depth of water acquisition by invading shrubs and resident herbs in a Sierra Nevada meadow. *Plant Soil* **2006**, *285*, 31–43. [CrossRef]
57. Wu, K.S. Study of Hydraulic Lift in *Zygophyllum xanthoxylum* of Eremophytes. Master's Thesis, Lanzhou University, Lanzhou, China, 2010.
58. Zhu, Y.J.; Jia, Z.Q. Water source of *Haloxylon ammodendron* plantations in autumn at the southeast edge of Badain Jaran Desert. *Sci. Silvae Sin.* **2012**, *48*, 1–5.

Article

Photosynthetic Responses of Two Woody Halophyte Species to Saline Groundwater Irrigation in the Taklimakan Desert

Jiao Liu [1,2], Ying Zhao [1,2,*], Tanveer Ali Sial [1,3], Haidong Liu [2], Yongdong Wang [2] and Jianguo Zhang [1,*]

1 Key Laboratory of Plant Nutrition and the Agri-Environment in Northwest China, Ministry of Agriculture, Northwest A&F University, Xianyang 712100, China; liuxy0803@163.com (J.L.); alisial@nwafu.edu.cn (T.A.S.)
2 Xinjiang Institute of Ecology and Geography, Chinese Academy of Sciences, Urumqi 830011, China; liu_hducas@163.com (H.L.); wangyd@ms.xjb.ac.cn (Y.W.)
3 Department of Soil Science, Sindh Agriculture University, Tandojam 70060, Pakistan
* Correspondence: yzhaosoils@gmail.com (Y.Z.); zhangjianguo21@nwafu.edu.cn (J.Z.)

Abstract: The study of plant photosynthesis under different degrees of drought stress can provide a deeper understanding of the mechanism of plant drought resistance. In the Taklimakan Desert, saline groundwater is the only local water source with regard to shelterbelt construction and determines plant growth and photosynthetic changes. In this study, daily dynamics of gas exchange parameters and their responses to photosynthetic photon flux density at three irrigation levels (W1 = 17.5, W2 = 25, W3 = 35 mm) were measured for two main species, i.e., *Calligonum mongolicum* (C) and *Haloxylon ammodendron* (H). *H* was better adapted to drought stress than *C*. Net photosynthetic rate (P_N) was mainly related to soil water status in the main root system activity layer. In July, the daily variations of P_N and transpiration (T_r) for *C* were higher than *H*. *C* increased water use efficiency (WUE) with increases in P_N, while *H* decreased T_r to obtain a higher WUE. Either *C* or *H*, drought reduced the low light and metabolic capacity, and thus decreased the light adaptability and photosynthesis potential. We suggest a prerequisite understanding of physiological mechanisms and possible plant morphological adjustments required to adapt plant species to desert drought conditions.

Keywords: desert plant; photosynthesis; drought stress; drip irrigation

Citation: Liu, J.; Zhao, Y.; Sial, T.A.; Liu, H.; Wang, Y.; Zhang, J. Photosynthetic Responses of Two Woody Halophyte Species to Saline Groundwater Irrigation in the Taklimakan Desert. *Water* **2022**, *14*, 1385. https://doi.org/10.3390/w14091385

Academic Editor: Alejandro Pérez-Pastor

Received: 18 March 2022
Accepted: 21 April 2022
Published: 24 April 2022

1. Introduction

Desertification is a global concern and vegetation restoration has been regarded as an effective ecological measure for desertification control. However, water deficiency is one of the major limitations that affect the physiological processes and ecological adaptability of plants in arid regions of the world [1,2]. Desert plants grow in a harsh environment with high temperatures and are normally exposed to drought, which seriously influences the distribution and growth of plants. However, desert plants habitually reduce water loss and alleviate high-radiation damage to their photosynthetic apparatus through specific morphological and physiological features [3]. It is well known that desert plants can tolerate adversity and adapt at different growth stages. There are also significant differences between plant species. It is crucial to develop a better understanding of the adaptation capabilities of plants to their environments and find ways to facilitate those adaptations.

Photosynthesis is sensitive to environmental changes and may be used to estimate plant adaptation to their habitats. As the basic unit of physiological metabolism and matter accumulation, photosynthesis is an important link to plant growth and metabolism [4,5]. Photosynthetic tissue adapts to high external radiation by protecting the assimilation process through diurnal adjustments in photochemical and non-photochemical processes [6]. Numerous studies have indicated that desert shrubs have evolved unique physiological and morphological characteristics in adapting to the harsh environment and exhibit a higher tolerance to water shortage than other plants [7]. However, their metabolic activities are still vulnerable to variations in water availability [8–10]. This vulnerability is due to the main physiological parameters of desert plants, such as stomatal conductance, photosynthetic rate and transpiration rate, which are variably inhibited during times of water shortage [11–15].

The Taklimakan Desert is located in northwest China and is the second-largest shifting sand desert in the world. It is an extremely harsh environment where most areas are covered by shifting sand. The Taklimakan Desert Highway Shelterbelt was constructed in 2003 to protect the highway and is irrigated with underground water. Three main plant species are grown in the shelterbelt, *Haloxylon ammodendron* (*H*), *Calligonum mongolicum* (C) and *Tamarix ramosissima*. These plant species are highly salt-tolerant and drought-resistant with excellent windbreak and sand fixation properties. However, it has long been speculated that these plant species may suffer from water and salt stresses [16]. The physiological characteristics of *C* and *H* and their photosynthetic organs conform with Kranz anatomy and are typical C4 plants with carbon isotopic ratios ($\delta^{13}C$) of -15%, CO_2 compensation points of less than 5 $\mu mol\ mol^{-1}$, light saturation points (LSP) of higher than 1600 $\mu mol\ m^{-2}\ s^{-1}$ and elevated photosynthetic capacities combined with high water use efficiencies [17]. Current concerns about the relationship between plant-available water and plants under drought stress focus mainly on the photosynthesis of natural vegetation in dry climates [17–20]. There is limited research on the photosynthetic characteristics of artificial forests under saltwater irrigation.

The primary problem encountered when trying to establish and support vegetation in desert ecosystems is the availability of good quality water. In the Taklimakan Desert, groundwater is the only local water source with regard to shelterbelt construction. However, the solutes in groundwater may increase soil salt contents and affect plant growth and survival in irrigated areas [21]. When exposed to salt stress, plant growth may be influenced by the osmotic effects on water uptake. The irrigation levels influence physiological activity, individual morphology and even the long-term adaptive strategies of desert plants. Under these circumstances, we tested how plants adapt to different soil moisture statuses and the impacts of different irrigation levels on *C* and *H*. We considered that the plant physiological response to irrigation levels should be closely related to soil moisture contents and that there may be variation between different species. Our objectives were to investigate: (1) the effects of irrigation levels on the plants and (2) the photosynthesis adaptations of two plant species in response to drought stress and, more specifically, the differences between them. Only when careful choices are made concerning the salinity of available irrigation water and the salinity tolerance of appropriate plant species can vegetation restoration promise sustainable development.

2. Materials and Methods

2.1. Environmental Conditions of the Study Area

The study was conducted at the Taklimakan Desert Research Station, Chinese Academy of Sciences (CAS). The station is located in the middle of the Taklimakan Desert (Tazhong), China (39°00′ N and 83°40′ E) at an altitude of 1099 m (Figure 1). The climate is an arid desert climate with an average annual precipitation of <50 mm and average annual evaporation of >3000 mm [22]. According to the Tazhong weather station's records, the average temperature is 12.4 °C, but the maximum and minimum temperatures may reach 45.6 °C and −22.2 °C, respectively. The soil is aeolian sand with low nutrient contents, poor moisture capacity and limited fertilizer retention. The salt content ranges from 1.26 to 1.63 $g\ kg^{-1}$, and the pH ranges from 8 to 9 [23]. The Taklimakan Desert Forest is an artificial plantation where *H. ammodendron* and *C. mongolicum* are the main tree species.

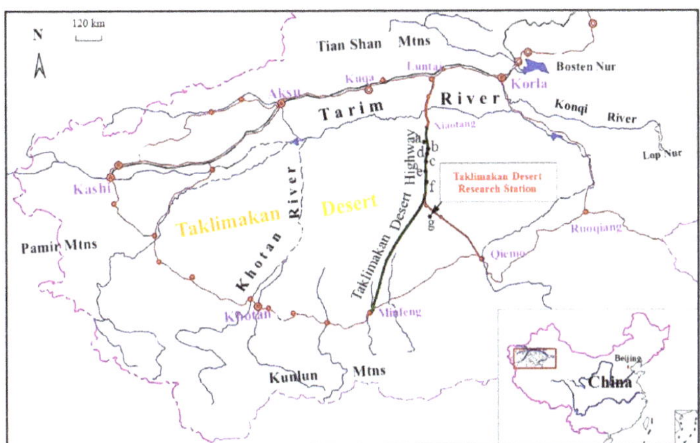

Figure 1. Taklimakan Desert Highway and the monitoring sites.

2.2. *Difference in Irrigation Management*

We selected six rows of *H* and *C* in the current study in 2016. All plants were the same age (8 years old) and were planted in the same field. Each row had one type of plant and the interval between adjacent rows was greater than 5 m to avoid water exchange between the different rows. Twenty trees were planted in each row and the distance between each tree was 1 m. Drip irrigation pipes were laid close to the trunk, the water outlet holes were spaced with 1 m, and the drip flow rate was 1.5 L/h. The specific field configuration is shown in Figure 2. All *H* and *C* trees were irrigated with saline groundwater with a 10-day time interval through a drip irrigation system with three irrigation levels of 17.5 mm (W1), 25 mm (W2) and 35 mm (W3), and all other basic agronomical practices were the same. In September and October, the time interval of irrigation increased so that plants were irrigated with groundwater every 15 days. Plants were not irrigated from November to February. The salt concentration of the groundwater was 4.02 g L^{-1}.

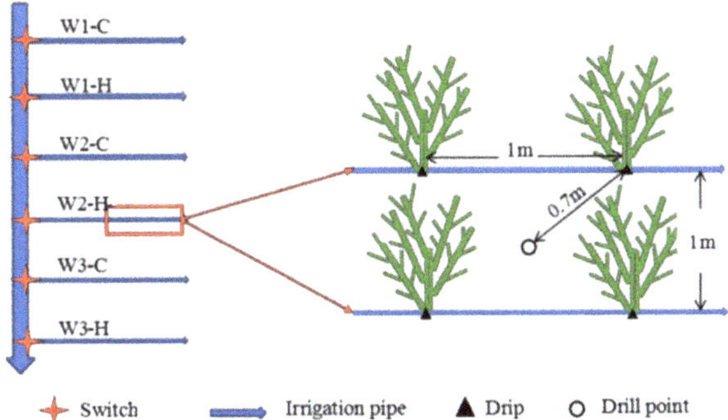

Figure 2. Schematic diagram of field experiment design.

2.3. *Soil Physical and Chemical Properties*

The soil moisture content was measured using the oven-dry method. Three soil samples were collected from each plot to determine the soil water content using oven-drying and weighing methods. Each soil sample was collected at a depth of 0–200 cm during photosynthesis measurement.

The soil was shifting Aeolian sand with a sandy texture. The salt content was 1.26 g kg^{-1} and it had a slightly alkaline reaction with a pH value of 7.69, a bulk soil density of 1.63 g cm^{-3}, field capacity of 21.27%, a soil organic matter content of 0.94 g kg^{-1}, a total soil nitrogen content of 0.03 g kg^{-1}, a total soil phosphorus content of 0.40 g kg^{-1} and a total soil potassium content of 16.62 g kg^{-1}.

2.4. Photosynthesis Measurements

The daily gas exchange dynamics and the responses to photosynthetic photon flux density (PPFD) were measured on typically sunny days during the same growth stage using an LI–6400 Portable Photosynthetic System (LI-COR, Lincoln, NE, USA). To investigate the drought-stress responses of the two shrub species when combined with high temperatures, we measured the daily gas exchange dynamics each day in July (Beijing time 9:00–20:00), and took measurements at intervals of two hours in each leaf. Sunnyside and undamaged assimilating shoots blade at the middle height of trees were chosen to measure. Four assimilating shoots were selected in each treatment, which were recorded five times for each assimilating shoot. The measurements were made when the sun was directly overhead to have a consistent sun angle. To keep the leaf chambers of the assimilating shoots closed during measurement, we marked the assimilating shoots with colored tape [17]. The same assimilating shoots were measured all day.

Photosynthetic photon flux density (PPFD) was measured using an LI–6400 Portable Photosynthetic System (LI–COR, Lincoln, NE, USA) from July to September to compare seasonal differences in 2016. According to the high temperature and strong radiation in the desert, we set a constant temperature of 30 °C, and the reference CO_2 concentration was equal to the environmental CO_2 concentration (380 mmol mol^{-1}). The light source was provided via an RGB Light, and the intensity ranged from 2200 to 0 mol m^{-2} s^{-1} for which the gradient gradually decreases in 12 grades. The data were recorded with the instrument's automatic observation programs, and each measurement was finished once.

After measuring photosynthesis, the leaves were taken off, and we took photos to calculate their areas with Photoshop, and then the photosynthetic parameters were recalculated.

The light compensation point (LCP) and apparent quantum yield (AQE) were obtained by fitting a linear regression of net photosynthetic rate (P_N) against PPFD ($\leq$200 μmol m^{-2} s^{-1}). The light saturation point (LSP), dark respiration rate (R_d) and maximum net photosynthetic rate (P_{max}) were obtained by fitting a Farqhuar non-linear hyperbolic model of P_N against PPFD ($\geq$200 μmol m^{-2} s^{-1}) [3,24].

$$P_N = \frac{\Phi * Q + P_{max} - \sqrt{(\Phi * Q + P_{max})^2 - 4 * \Phi * Q * k * P_{max}}}{2k} - R_d \quad (1)$$

where P_N is the net photosynthetic rate; Φ is apparent quantum yield; Q is effective photosynthetic radiation incident to the leaf; P_{max} is the maximum net photosynthetic rate; R_d is dark respiration rate; and k is the curved angle of the light response curve. φ

Plant water use efficiency (i.e., instantaneous; CO_2 assimilation/transpiration) is calculated as:

$$WUE = P_N/E \quad (2)$$

P_N is the net photosynthetic rate (μmol CO_2), and E is transpiration (mol H_2O).

2.5. Statistical Analyses

The statistical analyses were conducted using IBM SPSS Statistics (19.0, Software, USA) and figures were created using Origin 2016. The results for the parameters are presented as t-tests arranged according to photosynthetic rates under the different irrigation levels of the two woody plant species using IBM SPSS Statistics software 19.0. Multiple comparisons of various levels were performed using Duncan's New Multiple Range Test.

3. Results

3.1. Soil Moisture Content

Soil water content varied greatly between months. Overall, the water content of the shallow soil layer (0–100 cm) in July, August and September was greater than that of the deep soil layer (100–200 cm). The maximum soil water content was 40–60 cm, the average soil water content of 0–200 cm in July was the lowest, and that in September was the highest.

Differences between months of the same treatment were tested by paired-samples T test, which showed that all treated soil water content was insignificant between August and September, while differences between July and August and between July and September reached significant or extremely significant levels ($p < 0.01$).

In July, August and September, the soil water content of 0–100 cm layer under W2 treatment was less than W3 and W1 for C, and the content under W3 was the largest for *H*. The soil water content under W2 treatment was the lowest in July and August, and the soil water content under W1 treatment was the lowest in September for *H* and *C* (Figure 3).

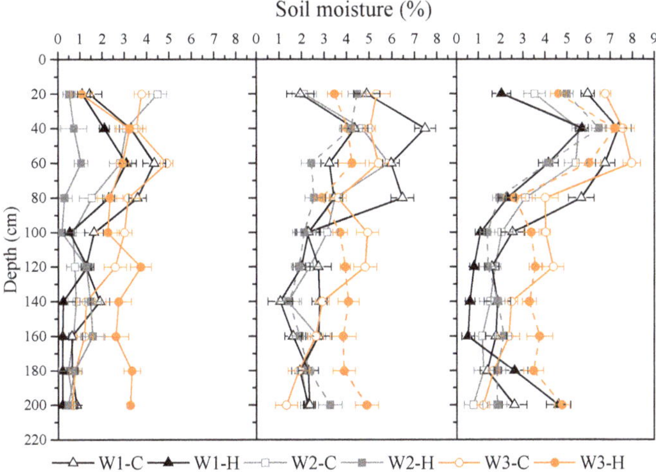

Figure 3. The soil moisture content in the low (W1), medium (W2) and high (W3) irrigation amounts during July, August and September of 2016. The C and H present *C. mongolicum* and *H. ammodendron*, respectively, same in the figures blow.

3.2. Daily Changes in the Meteorological Factors for the Different Months

Figure 4 indicated that the photosynthetic photon flux density (PPFD) increased antemeridian, reaching the highest value at noon, maintaining the maximum value until 15:30, and declining until sunset. The maximum PPFD was recorded in July, where the whole day maximum was 2112 μmol m^{-2} s^{-1} and was 1825 μmol m^{-2} s^{-1} in September. There was a trend of increasing air temperatures (T_a) throughout the experiment. A higher T_a was recorded in July than in August and September. A T_a greater than 30 °C was recorded after 10:30 and the maximum value was recorded at 17:30 at 40.1 °C in July. The mean T_a values in August and September between 8:30 and 20:00 were 25.58 °C and 24.54 °C, respectively, and there was no significant difference between the two months after 11:30.

The relative humidity (RH) declined from 8:30 to 20:00 during the day. The highest values were recorded in August and the lowest in July, and the mean values for July, August and September between 8:30 and 20:00 were 12.15%, 49.67% and 27.11%, respectively.

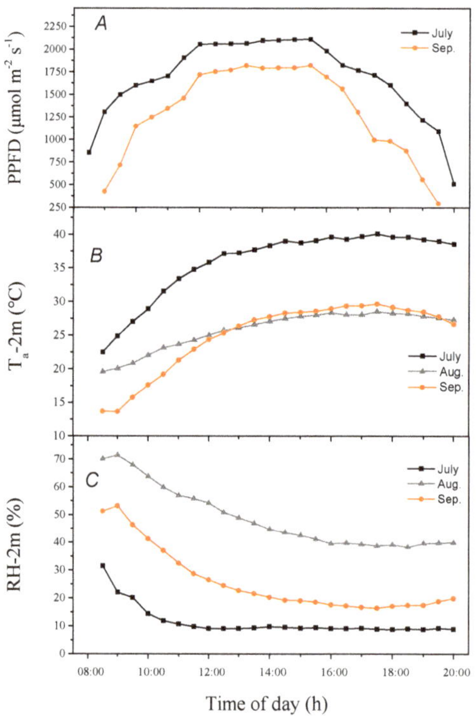

Figure 4. Diurnal changes in the photon flux density (PPFD) (**A**), air temperature (Ta) (**B**) and air relative humidity (RH) (**C**) for July, August and September in 2016.

3.3. Diurnal Changes of the Photosynthetic Rate under the Different Irrigation Levels

At the plant growth stage, we measured plant height and crown width under different treatments, and the analysis showed that there was no significant difference in the increase of plant height and crown width among all treatments. Different irrigation treatments had limited effects on plant height and crown width.

The diurnal differences in the net photosynthetic rates (P_N) of *C* and *H* under different irrigation levels are shown in Figure 5. The daily variations of P_N for *C* in the W1 treatment showed a bimodal curve and displayed an obvious "midday depression" phenomenon. However, only a slight "midday depression" phenomenon was observed in the W2 and W3 treatments. The depression of P_N for *C* in the W1, W2 and W3 treatments appeared around 14:00. For the daily mean P_N (from 9:00–19:00), the highest value was recorded in W1 (21.27 μmol (CO_2) m^{-2} s^{-1}), followed by the W3 (14.86 μmol (CO_2) m^{-2} s^{-1}) and W2 (13.80 μmol (CO_2) m^{-2} s^{-1}) treatments.

In contrast, P_N was generally lower in *H* than *C*. The daily variations of P_N for *H* in the W1 treatment showed a unimodal pattern with a maximum value around noon. The P_N values in the W2 and W3 treatments displayed a bimodal curve. A depression occurred around 13:00 h in the W2 treatment and around 15:00 h in the W3 treatment. The daily mean P_N (from 9:00 to 19:00) was highest in the W3 treatment with a maximum value of 11.61 μmol (CO_2) m^{-2} s^{-1}, followed by the W1 (7.85 μmol (CO_2) m^{-2} s^{-1}) treatment with a maximum value 10.25 μmol (CO_2) m^{-2} s^{-1} and the W2 (6.56 μmol (CO_2) m^{-2} s^{-1}) treatment with maximum value 8.98 μmol (CO_2) m^{-2} s^{-1}. The daily mean P_N of *C* was significantly higher in the W3 treatment than in the W2 treatment ($p < 0.05$). There were no statistically significant differences in the daily mean P_N values between the W1 and W2 treatments or between the W1 and W3 treatments ($p > 0.05$).

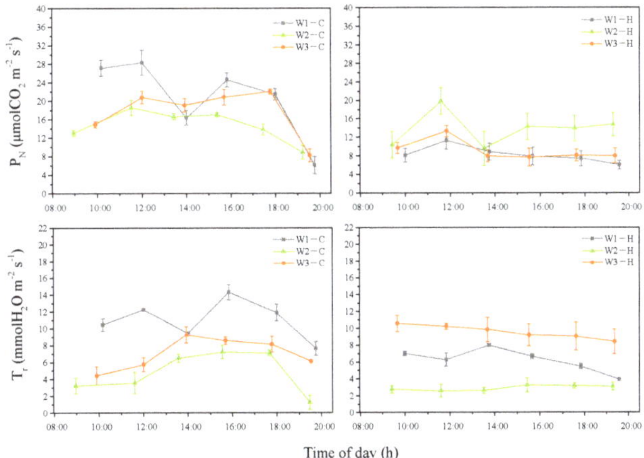

Figure 5. Diurnal changes in the net photosynthetic rate (P_N) and transpiration (T_r) of *C. mongolicum* (*C*) and *H. ammodendron* (*H*) in the low (W1), medium (W2) and high (W3) irrigation amounts.

Transpiration (T_r) for the different irrigation levels generally exhibited a similar pattern to net photosynthesis (Figure 3). The diurnal variations of T_r for *C* were unimodal and reached a peak at 13:30 h in the W2 and W3 treatments. On the contrary, the T_r values displayed a bimodal distribution in the W1 treatment and reached peaks at 12:00 and 16:00 h. For daily mean T_r (from 9:00 to 19:00) the highest values were obtained in the W1 treatment (13.25 mmol (H_2O) m^{-2} s^{-1} and 16.51 mmol (H_2O) m^{-2} s^{-1}), followed by the W3 treatment (8.97 mmol (H_2O) m^{-2} s^{-1} and 11.03 mmol (H_2O) m^{-2} s^{-1}) and the W2 treatment (7.24 mmol (H_2O) m^{-2} s^{-1} and 9.34 mmol (H_2O) m^{-2} s^{-1}. The diurnal variations of T_r for *H. ammodendron* were unimodal and reached a peak at 15:00 h in the W1 treatment. The T_r values decreased before 11:30 am and increased to reach a peak at 14:00 until sunset in the W2 and W3 treatments. The daily mean highest T_r was recorded from 9:00 to 19:00 in the W1 treatment (6.23 mmol (H_2O) m^{-2} s^{-1}), followed by the W3 (5.24 mmol (H_2O) m^{-2} s^{-1}) and W2 (2.93 mmol (H_2O) m^{-2} s^{-1}) treatments.

In all treatments, the diurnal changes of intercellular CO_2 concentration (C_i) of *C* and *H* exhibited a W-shaped response curve (Figure 6). The diurnal changes of L_S were opposite to that of C_i.

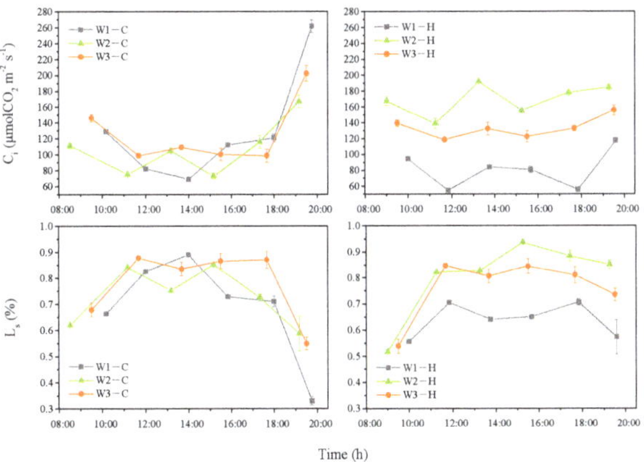

Figure 6. Differences in the intercellular CO_2 concentrations (C_i) and stomatal limitations (L_s) of *C. mongolicum* (*C*) and *H. ammodendron* (*H*) in the low (W1), medium (W2) and high (W3) irrigation amounts.

The factors influencing the decline of net photosynthetic rate of plants can be divided into stomatal limiting factors and non-stomatal limiting factors. C_i decreased and Ls increased when P_N decreased in W1-C treatment from 12:00 to 14:00, W1-H treatment from 14:00 to 18:00 and W3-H treatment from 14:00 to 16:00, indicating controls by stomatal limiting factors. In contrast, during other periods, P_N decreases were controlled by non-stomatal limiting factors.

3.4. The Response of P_N to Photosynthetic Photon Flux Density (PPFD) under Different Irrigation Levels

The response of P_N to photosynthetic photon flux density (PPFD) under different irrigation levels during July, August and September is presented in Figure 7. The physiological parameters are further calculated according to the response curve of P_N to photosynthetic photon flux density (PPFD), which reflects the photosynthetic capacities of the plant (Table 1).

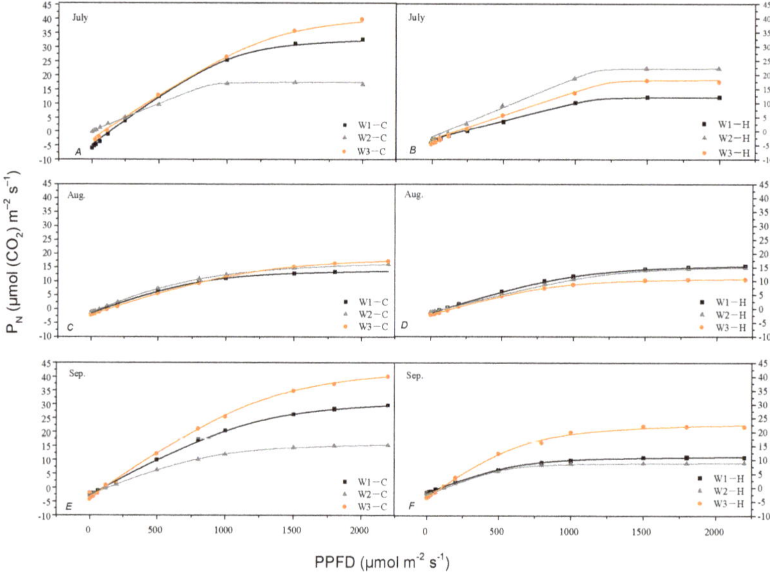

Figure 7. The photosynthetic light response curves (P_N) to photosynthetic photon flux density (PPFD) of *C. mongolicum* (C) and *H. ammodendron* (H) in the low (W1), medium (W2) and high (W3) irrigation amounts in July (**A**,**B**), August (**C**,**D**) and September (**E**,**F**).

The plant species obviously reacted differently during July, August and September due to the different responses of P_N to photosynthetic photon flux density (PPFD) with three irrigation levels. The photosynthesis of *C* irrigated with three irrigation levels was higher than *H* in July and September. However, in August, they were approximately equal in *C* and *H*. The light compensation point (LCP) values of *C* and *H* for the W2 treatment were lowest in July and August and highest in September. In comparing the W1 and W2 treatments, the light saturation point (LSP) of *C* in the W3 treatment was the highest in July and August. However, the LSP values of *C* and *H* were lowest in the W2 treatment irrigated with three irrigation levels in September. The light-saturated rate of CO_2 assimilation (P_{max}) and dark respiration rate (R_d) of *C* in the W3 treatment was obviously the highest values for those parameters during the three months. However, the P_{max} and R_d of *H* in the W3 treatment were higher than in the W1 and W2 treatments in September. The bivariate correlation test results compare the environmental variables and plant physiology characteristics in Table 2.

Plant photosynthesis is mainly related to PPFD and plant auto-regulatory mechanisms, but not to temperature and relative humidity.

Table 1. The light compensation point (LCP), light saturation point (LSP), the apparent quantum yield of CO_2 assimilation (Φ), dark respiration rate (R_d) and light-saturated rate of CO_2 assimilation (P_{max}) of *C* and *H* under the low (W1), medium (W2) and high (W3) irrigation amounts.

	LCP [μmol m^{-2} s^{-1}]			LSP [μmol m^{-2} s^{-1}]			Φ [mol mol^{-1}]			R_d [μmol m^{-2} s^{-1}]			P_{max} [μmol(CO_2) m^{-2} s^{-1}]		
Month	**July**	**Aug.**	**Sep.**	**July**	**Aug.**	**Sep.**	**July**	**Aug.**	**Sep.**	**July**	**Aug.**	**Sep.**	**July**	**Aug.**	**Sep.**
W1-C	146.87	90.68	108.65	1113.77	1131.84	1497.70	0.040	0.016	0.025	4.714	1.510	2.739	38.78	16.14	35.00
W2-C	49.52	71.01	138.94	1109.12	1195.87	1043.97	0.023	0.018	0.021	1.511	1.344	2.661	23.84	19.91	19.10
W3-C	214.07	128.94	100.56	1482.02	1528.18	1252.68	0.040	0.016	0.042	7.339	2.041	3.583	50.72	22.11	48.50
W1-H	244.13	97.25	95.74	1869.86	1213.77	811.68	0.013	0.016	0.019	6.036	1.574	1.804	21.78	18.31	13.39
W2-H	130.17	97.09	123.69	1096.40	1371.49	618.06	0.025	0.013	0.024	1.744	1.302	2.624	24.35	17.08	11.77
W3-H	169.02	163.86	100.51	1082.83	1281.61	871.18	0.023	0.012	0.036	2.615	2.038	3.570	21.11	13.41	27.67

Table 2. The bivariate correlation test results comparing the environmental variables and plant physiology characteristics (P_N—Net photosynthetic rate; T_r—Transpiration rate; C_i—Intercellular CO_2 concentration; L_s—stomatal limitation; g_s—Stomatal conductance; PPFD—Photosynthetic photo flux density; T_a—Air temperature; RH—Relative humidity; WUE—Water use efficiency; T_l—Leaf temperature; *C*—*C. mongolicum*; *H*—*H. ammodendron*) during July 2016.

	Plant Species	P_N	T_r	C_i	L_s	g_s	PPFD	T_a	RH	WUE	T_l
P_N	*C*	1.00									
	H	1.00									
T_r	*C*	0.76 **	1.00								
	H	0.80 **	1.00								
C_i	*C*	−0.66 **	−0.25	1.00							
	H	−0.34	−0.45	1.00							
L_s	*C*	0.65 **	0.24	−0.99 **	1.00						
	H	0.34	0.45	−.99 **	1.00						
g_s	*C*	0.69 **	0.72 **	−0.23	0.21	1.00					
	H	0.76 **	0.80 **	−0.44	0.44	1.00					
PPFD	*C*	0.66 **	0.51*	−0.69 **	0.69 **	0.13	1.00				
	H	0.29	0.22	−0.34	0. 34	−0.04	1.00				
T_a	*C*	0.04	0.20	−0.05	0.08	−0.44	0.42	1.00			
	H	−0.10	−0.07	−0.19	0.21	−0.37	0.49 **	1.00			
RH	*C*	−0.30	−0.31	0.15	−0.17	0.25	−0.54 *	−0.91 **	1.00		
	H	−0.01	0.06	0.28	−0.29	0.35	−0.55 *	−0.94 **	1.00		
WUE	*C*	0.55 *	−0.09	−0.73 **	0.73 **	0.12	0.41	−0.13	−0.13	1.00	
	H	−0.19	−0.69 **	0.41	−0.42	−0.51 *	−0.06	0.08	−0.01	1.00	
T_l	*C*	0.31	0.46	−0.19	0.21	−0.18	0.68	0.89 **	−0.87 **	−0.09	1.00
	H	−0.03	0.13	−0.28	0.30	−0.26	0.59 **	0.86 **	−0.82 **	−0.33	1.00

* Correlation is significant at the 0.05 level (2-tailed *t*-tests). ** Correlation is significant at the 0.01 level (2-tailed *t*-tests).

4. Discussion

4.1. The Limiting Factors Affecting Photosynthesis

Four major factors affect soil moisture regime in our studied system: precipitation, irrigation, crop evapotranspiration and soil evaporation. Soil moisture content under the same irrigation level was different in different seasons. In July, the soil moisture content within a 60–80 cm soil depth for the W2 treatment was lower than the W1 treatment. Plant roots are mainly concentrated in the 0–80 cm soil layer [25], and the plant transpiration in desert soil was below 60 cm, so these conditions may have resulted in the lower soil moisture contents in the W2 treatment. With the decrease of soil evaporation and plant transpiration in August and September, the soil water was replenished at the 0–80 cm depth. The irrigation levels should be adjusted by combining the environmental conditions, soil moisture content and water requirements of plants in different seasons.

Water is the primary limiting factor for plant growth in arid and semiarid regions. For desert plants, the compromise occurs in nature between restricting water loss through the stomata versus maintaining a high carbon gain, which depends on both stomatal and non-stomatal regulations [3]. The intercellular CO_2 concentration (C_i) should change in the same direction as the P_N if stomatal responses are dominant in response to perturbations to the assimilation rate. If the changes are opposite, the most crucial difference must have been in the mesophyll cells [26], caused by non-stomatal factors [7,18].

When plants are exposed to moderate or severe drought stress, the stomatal restriction is no longer the main reason for the decrease of photosynthesis and the role of non-stomatal restriction becomes more important [7]. In the present study, the changes in the P_N of *C* between 11:00 and 15:00 h in the W1 treatment and between 13:00 and 17:00 h in the W3 treatment were in the same direction as the intercellular CO_2 concentration (C_i) (Figure 6), which indicated that stomatal factors controlled the photosynthetic processes at these times. The C_i changed in the opposite direction as the P_N in the W2 treatment throughout the day, indicating that non-stomatal factors changed the photosynthetic processes. The C_i changed in the same direction as the P_N for *H* in the W1 treatment between 13:00 and 17:00 h and the W3 treatment between 11:00 and 13:00 h, which indicated that stomatal factors changed the photosynthetic processes during this period. The photosynthetic processes of *C* and *H* in the W2 treatments were restricted to non-stomatal factors through the day and changes in soil moisture between 60–80 cm. These phenomena indicated that P_N was mainly related to the soil water status in the main root system activity layer. The P_N values for *C* and *H* were mainly limited by non-stomatal factors under severe water stress [27].

A "midday depression" is one of the ways for plants to adapt to a dry environment through evolution. *C* and *H* are typical desert plants and the diurnal variations of P_N and T_r for *C* in the W1 treatment presented an obvious "midday depression" phenomenon, but the W2 and W3 treatments presented a slight "midday depression" phenomenon. This indicates that *C* was under serious water stress, resulting in partial stomata closure to reduce transpiration. In contrast, the diurnal variations of P_N for *H* in the W2 treatment exhibited an obvious "midday depression" phenomenon because the curves of P_N for *H* in the W1 and W3 treatments were unimodal and there were no symptoms that *H* suffered water stress in the W1 treatment.

In this study, the daily average values of P_N and T_r for *C* were obviously higher than for *H* under the same irrigation levels. The *C* in the W1 treatment suffered from water stress, and *H* in the W1 treatment did not demonstrate drought stress, which indicated that *H* is more drought-tolerant than *C*, and a lower T_r benefits water preservers.

Many studies have suggested that desert plants generally have high water use efficiency, especially under water stresses; high water use efficiency is usually an effective way to resist drought stress [28–30]. The daily water use efficiency (WUE) of *C* was 1.61, 1.91 and 1.66 $\mu molCO_2\ mmolH_2O^{-1}$ for the W1, W2 and W3 treatments and for *H* it was 1.25, 2.20 and 1.24 $\mu molCO_2\ mmolH_2O^{-1}$, respectively. For the two plant species, the WUE was higher in the W2 treatment. This indicated that light water stress could improve the WUE of *C* and *H*, but WUE cannot be improved when plants are water deficient. This result may occur because water deficiency inhibits the plant's photosynthetic rate. Many stressful climate features include high temperature, high radiation and high evaporation in arid desert regions. The plant must maintain a certain transpiration rate to reduce the temperature and resist burns on its leaves [31].

Many environmental factors such as high light intensity, temperature and low soil moisture cause photo inhibition [32–34]. Our results indicated that environmental factors such as PPFD, T_a and RH changed from morning to night, as indicated by the bivariate correlation test results comparing the environmental variables and plant physiology characteristics presented in Table 2. g_s is a major determinant of the photosynthetic rate under well-watered conditions and non-stomatal limitations during drought conditions. P_N and T_r for *C* and *H* were significantly correlated with g_s. This result is consistent with previous findings [35–37]. The P_N of *C* was significantly correlated with PPFD, C_i, L_s and g_s. However, the P_N of *H* was correlated with g_s. The WUE of *C* was correlated with P_N, C_i and L_s, while the WUE of *H* was correlated with T_r and g_s. All of the results showed that the photosynthetic organs of *C* are more sensitive to the PPFD changes than *H*. *C* can adjust the relationship between respiration and photosynthesis, resulting in increases in P_N and WUE. However, *H* can adjust the g_s to decrease T_r, thus resulting in a higher WUE.

4.2. Plants' Adaptation Strategy to Drought Stress

The photosynthesis–light response curve helps estimate the adaptation of plants to their habitats. The light compensation point (LCP), light saturation point (LSP), apparent quantum yield of CO_2 assimilation (Φ), dark respiration rate (R_d) and light saturated rate of CO_2 assimilation (P_{max}) of *C* and *H* were different for the all of the irrigation treatments during the July, August and September throughout the experimental period. These physiological parameters reflect the adaptability of plants to environmental changes (Table 1).

The LCP reflects the ability of plants to use weak light, LSP demonstrates the ability of plants to use strong light and different values between LSP and LCP reproduce the plants' light adaptation. In the present study, the LCP of *C* and *H* in the W2 treatment was lower than the W1 and W3 treatments from July to August but was higher in September, and soil evaporation was stronger in July and August. Plants were stressed in two months. In September, soil moisture was higher, indicating that *C* and *H* can decrease LCP to adapt to drought stress [38]. The phenomenon of photosynthesis domestication appeared. The LSP of *C* was greater in the W3 treatment in July and August, but the LSP of *H* was not regular for the three irrigation levels. The different LSP and LCP values of *C* in the W3 treatment were higher from July to August. The values of *C* and *H* in W2 were lower than the W1 and W3 treatments in September, which indicated that *C* might make better use of strong light and show better light adaptation under better water conditions. Drought stress might improve the light adaptation of *H* and cause *H* to adapt to drought stresses.

R_d for *C* and *H* was higher in the W3 treatment than in the W1 and W2 treatments during the three months we tested. This indicated that drought stress might slow the plants' metabolic capacity. This is a protective response of plants to drought stress. A decrease of P_N under drought stress led to a shortage of photosynthetic matter in the body. A decrease of R_d reduces the consumption of organic matter due to breathing, and it ensures that plants have a relatively large number of photosynthetic substances to construct competitive organs, such as the main root, lateral root, etc. Drought stress might relate the activity of key enzymes associated with changes in respiration concentration [39], and may also be caused by long-term severe drought stress destruction of plant cell structures related to respiration.

C in the W3 treatment had a higher P_{max} than W1 and W2 treatments. However, *H* in the W3 treatment had a lower P_{max} than the W1 and W2 treatments. The photosynthesis potential of *C* slowed down during drought stress, but drought inspired *H* photosynthesis potential. Furthermore, studies have shown that the plant shoots can improve their resistants (such as drought), making their newly developed photosynthesis potential as an effective means to increase yield [40–42]. The dark respiration rate for shoots of *H* is mainly affected by the effective photosynthetic radiation, air temperature, and relative humidity. However, photosynthesis rate is less affected by the environment, making its strong resistance [42]. At present, the contribution of shoot photosynthesis to drought tolerance cannot be ignored, and the response of shoot photosynthesis to drought needs to be further explored.

Finally, in addition to the two species selected in this study, several other species, particularly Tamarix ramosissima, are widely distributed in the region. As there are significant photosynthesis differences among plants and groundwater salinities, species combinations at different shelterbelt engineering areas differ. It is necessary to research more species with regard to plant species survival characteristics by using water-controlled experiments for a complete understanding of physiological mechanisms that is difficult to obtain in natural conditions due to the lack of a vast water range. More suitable plants should be selected, and sustainable irrigation technologies should be developed in this area and beyond.

5. Conclusions

This study demonstrated that two species (*C* and *H*) have salt resistance capabilities that maintain normal leaf scale photosynthesis during topsoil water stress conditions. Their adaptive strategies are eco-physiologically different at the species level. The soil water availability determines the plant response, acclimation and adaptation in this mobile sand desert. P_N was mainly related to soil water status in the main root system activity layer. The daily variations of P_N and T_r for *C* were higher than that for *H* in July. Compared to *C*, *H* demonstrated a stronger capability for drought resistance. *C* is more sensitive to changes in PPFD than *H* as *C* increased WUE through increased P_N, but *H* decreased T_r to obtain a higher WUE. Drought reduced the use of weak light and metabolic capacity of *C* and *H*, and decreased the light adaptability and photosynthesis potential of *C* instead of *H*. To evaluate the drought hardiness of plant species survival for extreme desert conditions, which are often difficult to obtain in natural conditions, our water-controlled studies are beneficial for a complete understanding of physiological mechanisms and possible plant morphological adjustments.

Author Contributions: Conceptualization and methodology, J.L. and Y.Z.; formal analysis, J.L.; investigation, J.L., H.L. and T.A.S.; resources, Y.W.; writing—original draft preparation, J.L.; writing—review and editing, J.L., Y.Z. and J.Z.; supervision, Y.Z. and J.Z.; project administration, Y.Z., Y.W. and J.Z. All authors have read and agreed to the published version of the manuscript.

Funding: This research was supported by the National Natural Science Foundation of China (41977009, 41877541, 41471222) and the National Talents Project (Y472241001).

Conflicts of Interest: The authors declare no conflict of interest.

Abbreviations

AQE—apparent quantum yield; C_i—intercellular CO_2 concentration; g_s—stomatal conductance; k —the curved angle of the light response curve; L_s—stomatal limitations; LCP—light compensation point; LSP—light saturation point; RH—air relative humidity; T_a—air temperature; T_l—leaf temperature; T_r–transpiration; P_N—net photosynthetic rate; P_{max}—maximum net photosynthetic rate light; Q—effective photosynthetic radiation incident to the leaf; R_d—dark respiration rate; Φ—apparent quantum yield; WUE—water use efficiency.

References

1. Ivans, S.; Hipps, L.; Leffler, A.J.; Ivans, C.Y. Response of water vapor and CO_2 fluxes in semiarid lands to seasonal and intermittent precipitation pulses. *J. Hydrometeorol.* **2006**, *7*, 995–1010. [CrossRef]
2. Resco, V.; Ignace, D.D.; Sun, W.; Huxman, T.E.; Weltzin, J.F.; Williams, D.G. Chlorophyll fluorescence, predawn water potential and photosynthesis in precipitation pulse-driven ecosystems—Implications for ecological studies. *Funct. Ecol.* **2008**, *22*, 479–483. [CrossRef]
3. Su, P.; Cheng, G.; Yan, Q.; Liu, X. Photosynthetic regulation of C_4 desert plant Haloxylon ammodendron under drought stress. *Plant Growth Regul.* **2007**, *51*, 139–147. [CrossRef]
4. Munns, R. Comparative physiology of salt and water stress. *Plant Cell Environ.* **2002**, *25*, 239–250. [CrossRef] [PubMed]
5. Su, P.; Yan, Q. Photosynthetic characteristics of C_4 desert species *Haloxylon ammodendron* and *Calligonum mongolicm* under different moisture conditions. *Acta Ecol. Sin.* **2006**, *26*, 75–82.
6. Franco, A.C.; Lüttge, U. Midday depression in savanna trees: Coordinated adjustments in photochemical efficiency, photorespiration, CO_2 assimilation and water use efficiency. *Oecologia* **2002**, *131*, 356–365. [CrossRef]
7. Lawlor, D.W.; Cornic, G. Photosynthetic carbon assimilation and associated metabolism in relation to water deficits in higher plants. *Plant Cell Environ.* **2002**, *25*, 275–294. [CrossRef]
8. Horton, J.L.; Kolb, T.E.; Hart, S.C. Leaf gas exchange characteristics differ among Sonoran Desert riparian tree species. *Tree Physiol.* **2001**, *21*, 233–241. [CrossRef]
9. Yang, Q.; Zhao, W. Responses of leaf stomatal conductance and gas exchange of Haloxylon ammodendron to typical precipition event in the Hexi Corridor. *J. Desert Res.* **2014**, *34*, 419–425.

10. Zan, D.; Zhuang, L. Drought resistance analysis based on anatomical structure of assimilating shoots of Haloxylon ammodendron from Xinjiang. *J. Arid Land Res. Environ.* **2017**, *31*, 146–152.
11. Levitt, J. Chilling, Freezing, and High Temperature Stresses. In *Responses of Plants to Environmental Stress*; Academic Press: New York, NY, USA, 1980; Volume 1.
12. Smith, S.D.; Devitt, D.A.; Sala, A.; Cleverly, J.R.; Busch, D.E. Water relations of riparian plants from warm desert regions. *Wetlands* **1998**, *18*, 687–696. [CrossRef]
13. Wang, X.; Luo, N.; Shan, H.; Wang, X. Responses characteristics of C_4 desert shrubs in Minqin under drought stress. *Arid Land Geogr.* **2016**, *39*, 1025–1035.
14. Wang, F.; Liu, S.; Kang, C.; Delu, L.; Chen, Z.; Li, X. Effects of drought stress on photosynthesis and chlorophyll fluorescence characteristics of Picea mongolica. *J. Arid Land Resour. Environ.* **2017**, *31*, 142–147.
15. Yang, S.; Fan, J.; Sun, Y.; Li, C.; Xu, X.; Fan, J.; Wang, J.; Zhang, H.; Zhai, Z. Photosynthetic characteristics and response of Haloxylon ammodendron to drought stress in Hinterland of the Lop Nur. *Arid Zone Res.* **2018**, *35*, 379–386.
16. Wang, B.; Jin, M.; Nimmo, J.R.; Yang, L.; Wang, W. Estimating groundwater recharge in Hebei Plain, China under varying land use practices using tritium and bromide tracers. *J. Hydrol.* **2008**, *356*, 209–222. [CrossRef]
17. Su, P.; Liu, X.; Zhang, L.; Zhao, A.; Li, W.; Chen, H. Chen comparison of d 13C values and gas exchange of assimilating shoots of desert plants H. ammodendron and C. mongolicum with other plants. *Isr. J. Plant Sci.* **2004**, *52*, 87–97. [CrossRef]
18. Tezara, W.; Marín, O.; Rengifo, E.; Martínez, D. Photosynthesis and photoinhibition in two xerophytic shrubs during drought. *Photosynthetica* **2005**, *43*, 37–45. [CrossRef]
19. Maina, N.J.; Wang, Q. Seasonal Response of Chlorophyll a/b Ratio to Stress in a Typical Desert Species: Haloxylon ammodendron. *Arid Land Res. Manag.* **2015**, *29*, 321–334. [CrossRef]
20. Wu, Y.; Zheng, X.; Li, Y. Photosynthetic response of desert plants to small rainfall events in the Junggar Basin, northwest China. *Photosynthetica* **2016**, *54*, 3–11. [CrossRef]
21. Zhang, J.G.; Xu, X.; Lei, J.Q.; Sun, S.G.; Fan, J.L.; Li, S.; Gu, F.; Qiu, Y.Z.; Xu, B. The salt accumulation at the shifting aeolian sandy soil surface with high salinity groundwater drip irrigation in the hinterland of the Taklimakan Desert. *Chin. Sci. Bull.* **2008**, *53*, 63–70. [CrossRef]
22. Lei, J.; Li, S.; Fan, D.; Zhou, H. Classification and regionalization of the forming environment of windblown sand disasters along the Tarim Desert Highway. *Chin. Sci. Bull* **2008**, *53*, 1–7. [CrossRef]
23. Han, W.; Cao, L.; Yimitb, H.; Xu, X.W. Optimization of the saline groundwater irrigation system along the Tarim Desert Highway Ecological Shelterbelt Project in China. *Ecol. Eng.* **2012**, *40*, 108–112. [CrossRef]
24. Thornley, J.H.M. Instantaneous canopy photosynthesis: Analytical expressions for sun and shade leaves based on exponential light decay down the canopy and an acclimated non-rectangular hyperbola for leaf photosynthesis. *Ann. Bot.* **2002**, *89*, 451–458. [CrossRef] [PubMed]
25. Zhang, J.G.; Lei, J.Q.; Wang, Y.D.; Xu, X.W. Survival and growth of three afforestation species under high saline drip irrigation in the Taklimakan Desert. *Ecosphere* **2016**, *7*, e01285. [CrossRef]
26. Farquhar, G.D.; Sharkey, T.D. Stomatal conductance and photosynthesis. *Ann. Rev. Plant Physiol.* **1982**, *33*, 317–345. [CrossRef]
27. Yu, Q.; Li, D.; Zhao, M.; He, Y.; Ma, Q.; Guo, S. Photosynthetic characteristics of Chilopsis linearis under soil water stress. *Acta Bot. Boreali-Occident. Sin.* **2007**, *27*, 2531–2539.
28. Casper, B.B.; Forseth, I.N.; Wait, D.A. A stage-based study of drought response in Cryptantha flava (Boraginaceae): Gas exchange, water use efficiency, and whole plant performance. *Am. J. Bot.* **2006**, *93*, 978–987. [CrossRef]
29. Gong, J.R.; Zhao, A.F.; Huang, Y.M.; Zhang, X.S.; Zhang, C.L. Water relations, gas exchange, photochemical efficiency, and peroxidative stress of four plant species in the Heihe drainage basin of northern China. *Photosynthetica* **2006**, *44*, 355–364. [CrossRef]
30. Rouhi, V.; Samson, R.; Lemeur, R.; van Damme, P. Photosynthetic gas exchange characteristics in three different almond species during drought stress and subsequent recovery. *Environ. Exp. Bot.* **2007**, *59*, 117–129. [CrossRef]
31. Yan, H.; Zhang, X.; Xu, H. Photosynthetic characteristics responses of three plants to drought stress in Tarim Desert Highway Shelterbelt. *Acta Ecol. Sin.* **2010**, *30*, 2519–2528.
32. Escalona, J.M.; Flexas, J.; Medrano, H. Stomatal and no-stomatal limitation of photosynthesis under water stress in field-grown grapevines. *Austral J. Plant Physiol.* **1999**, *26*, 421–433. [CrossRef]
33. Xu, H.; Li, Y.; Xu, G.; Zhou, T. Ecophysiological response and morphological adjustment of two Central Asian desert shrubs towards variation in summer precipitation. *Plant Cell Environ.* **2007**, *30*, 399–409. [CrossRef] [PubMed]
34. Ayup, M.; Hao, X.; Chen, Y.; Li, W.; Su, R. Changes of xylem hydraulic efficiency and native embolism of Tamarix ramosissima Ledeb. seedlings under different drought stress conditions and after rewatering. *S. Afr. J. Bot.* **2011**, *78*, 75–82. [CrossRef]
35. Zhao, C.; Wei, X.; Yu, Q.; Deng, J.; Wang, G. Photosynthetic characteristics of Nitraria tangutorum and Haloxylon ammodendron in the ecotone between oasis and desert in Minqin, Region, Country. *Acta Ecol. Sin.* **2005**, *25*, 1908–1913.
36. Sugaya, N.; Kuzushima, K. Increased leaf photosynthesis caused by elevated stomatal conductance in a rice mutant deficient in SLAC1, a guard cell anion channel protein. *J. Exp. Bot.* **2012**, *63*, 5635–5644.
37. Tian, Y.; Taxiplat, T.; Xu, G. Gas Exchange of Haloxylon ammodendron and H. persicum. *Arid Zone Res.* **2014**, *31*, 542–549.
38. Yan, H.; Zhang, X.; Xu, H.; Yao, S. Responses of Calligonum arborescens photosynthesis to water atress in Tarim Highway Shelterbelt. *J. Desert Res.* **2007**, *27*, 460–465.

39. Kumari, N.; Batra, N.G.; Sharma, V. Photosynthetic performance and drought-induced changes in activity of antioxidative enzymes in different varieties of Vigna radiata. *Agric. Res.* **2018**, *7*, 1–9. [CrossRef]
40. Ávila-Lovera, E.; Zerpa, A.J.; Santiago, L.S. Stem photosynthesis and hydraulics are coordinated in desert plant species. *New Phytol.* **2017**, *216*, 1119–1129. [CrossRef] [PubMed]
41. Simkin, A.J.; Faralli, M.; Ramaoorthy, S.; Lawson, T. Photosynthesis in non-foliartissues: Implications for yield. *Plant J.* **2020**, *101*, 1001–1015. [CrossRef] [PubMed]
42. Feng, X.L.; Liu, R.; Ma, J.; Xu, Z.; Wang, Y.; Kong, L. Photosynthetic characteristics and influncing factors of Haloxylon persicun stems (different diameter classes) in Gurbantonggut desert. *Acta Ecol. Sin.* **2021**, *41*, 9784–9785.

Article

Effects of Irrigation Regimes on Soil Water Dynamics of Two Typical Woody Halophyte Species in Taklimakan Desert Highway Shelterbelt

Jiao Liu [1], Ying Zhao [2,3,*], Jianguo Zhang [1], Qiuli Hu [2] and Jie Xue [3,4,5,6,*]

1 College of Resources and Environment, Northwest Agriculture and Forestry University, Xianyang 712100, China; liuxy0803@163.com (J.L.); zhangjianguo21@nwafu.edu.cn (J.Z.)
2 College of Resources and Environmental Engineering, Ludong University, Yantai 264025, China; qiulihu@ldu.edu.cn
3 State Key Laboratory of Desert and Oasis Ecology, Xinjiang Institute of Ecology and Geography, Chinese Academy of Sciences, Urumqi 830011, China
4 Cele National Station of Observation and Research for Desert-Grassland Ecosystems, Cele 848300, China
5 Xinjiang Key Laboratory of Desert Plant Roots Ecology and Vegetation Restoration, Xinjiang Institute of Ecology and Geography, Chinese Academy of Sciences, Urumqi, 830011, China
6 University of Chinese Academy of Sciences, Beijing 100049, China
* Correspondence: yzhaosoils@gmail.com (Y.Z.); xuejie11@ms.xjb.ac.cn (J.X.); Tel./Fax: +86-0991-7885394 (J.X.)

Citation: Liu, J.; Zhao, Y.; Zhang, J.; Hu, Q.; Xue, J. Effects of Irrigation Regimes on Soil Water Dynamics of Two Typical Woody Halophyte Species in Taklimakan Desert Highway Shelterbelt. *Water* **2022**, *14*, 1908. https://doi.org/10.3390/w14121908

Academic Editor: Guido D'Urso

Received: 13 April 2022
Accepted: 10 June 2022
Published: 14 June 2022

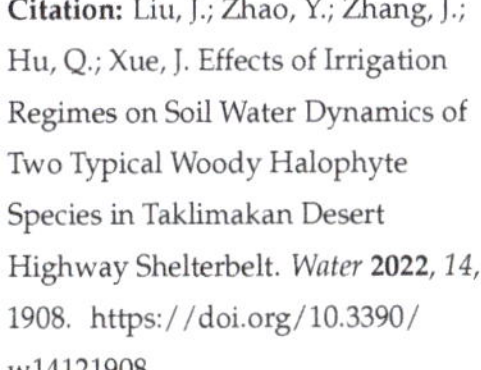

Abstract: Freshwater resources are in a shortage in arid regions worldwide, especially in extremely arid desert areas. To solve this problem, highly saline groundwater is used for drip irrigation of desert plants. Since more irrigation infiltrating into the deep soil cannot be absorbed and utilized by desert plants, it is crucial to determine optimal water-saving irrigation regimes. In this study, we examined the effects of irrigation regimes on the soil water dynamics of two typical woody halophyte species (*Haloxylon* and *Calligonum*), and quantified the irrigation intervals and periods based on a field test of precision irrigation control in the Taklimakan Desert Highway shelterbelt. Results showed that the change in soil moisture of two species in the shallow 0–60 cm layer could be divided into a rapid decline period (1–9 d), a slow decline period (9–19 d), and a relatively stable period (19–39 d) after irrigation. The decrease rate of soil moisture at the 0–60 cm depth was significantly higher than that at the 60–200 cm depth. The irrigation regime combining 35 mm irrigation with 10 days was beneficial to soil water storage and plant use with respect to *Calligonum*, while the irrigation regime combining 35 mm irrigation with 40 days was best for *Haloxylon*. Increasing the single irrigation amount and prolonging the irrigation period can further enable the more effective use of irrigation water. This study highlights that saline groundwater irrigation provides potential advantages for desert plants' survival under reasonable irrigation regimes.

Keywords: irrigation amount; irrigation periods; Taklimakan Desert Highway shelterbelt; soil water storage; woody halophyte; saline-tolerant plant

1. Introduction

Desertification is an environmental issue of global concern, especially in arid and semiarid regions. Artificial afforestation has been considered an effective ecological means for combating desertification in many arid desert regions worldwide [1,2]. Great challenges, however, have appeared when the afforestation is conducted in arid desert regions due to the lack of freshwater and extreme environmental conditions, including severe drought [1,3,4]. Due to less rainfall in arid desert regions, water scarcity has become a worldwide issue of increasing severity [1,5]. The lower-quality saline–alkaline groundwater is widely applied [6–8]. Unfortunately, saline water irrigation normally leads to greater salinity hazards to plant growth and survival in groundwater extraction [9,10]. Therefore, establishing a suitable irrigation regime for artificial vegetation growth and survival is crucial to saving groundwater utilization and reducing the salinity hazards.

The Taklimakan Desert, called the "Dead Sea", is the second-largest mobile desert in the world. To improve transportation for the exploitation of petroleum resources, the Taklimakan Desert Highway, the longest highway across a shifting desert in the world, was completed in 1995 [11]. To overcome the frequent sand burial of the highway, the Taklimakan Desert Highway shelterbelt was constructed through a biological engineering project in 2003 [12]. The mobile dunes on both sides of the highway were effectively stabilized by introducing drought- and salt-tolerant plants [13], such as *Haloxylon Bunge*, *Calligonum Linn*, and *Tamarix Linn*.

Drip irrigation from saline groundwater is one of the most efficient ways to support artificial shelterbelts [14]. Owing to the differences in the adaptability or adaptive strategies of drought- and salt-tolerant plants, irregular or insufficient irrigation will lead to different responses of plants to water stresses [14]. Therefore, it is crucial to determine suitable irrigation regimes to ensure plant survival in the drip irrigation process. *Haloxylon ammodendron* and *Calligonum mongolicunl* are the two main species in the Taklimakan Desert Highway shelterbelt. It is reported that moderate irrigation intervals are beneficial to the growth of the two species in the Taklimakan Desert Highway shelterbelt since they can save water and support the plants' water demands [14].

In recent years, many researchers have studied the soil water dynamics and irrigation regimes of desert plants. For example, Ding et al. reported that the soil moisture at 0–120 cm depth presents an apparent single-peak curve. The salt accumulation phenomenon is evident at 45–60 cm, and the salt content reaches 10–20 g kg^{-1} in the Taklimakan Desert [15]. Li et al. found that saltwater irrigation did not produce salt stress on the plant roots in the Taklimakan Desert [1]. *Haloxylon ammodendron*'s roots are mainly distributed between 20 and 80 cm, while the salt is mainly concentrated in the 0–20 cm surface layer. On the contrary, saline water irrigation is beneficial for increasing soil nutrients. Fu et al. pointed out that *Haloxylon ammodendron* mainly utilizes shallow soil moisture (20–40 cm) and deep soil moisture (100–350 cm) and underground water in May, but deep soil moisture (160–350 cm) and underground water in August in the southern edge of Gurbantunggut Desert [16]. Zhang et al. indicated that the structural heterogeneity of the soil layer has a retarding effect on water content [17]. The soil layers with more clay and silt particles are more prone to salt accumulation at the southern edge of the Gurbantunggut Desert [17].

The previous studies focused mainly on saline water irrigation and its influence on soil properties and plant growth [6,18,19]. However, the effects of saline water irrigation regimes on the soil water dynamics of desert plants have been ignored. Meanwhile, the number of irrigation intervals and periods of different desert plants remain unknown. This study aims to (1) examine the effects of irrigation regimes on the soil water dynamics of two typical woody halophyte species and (2) quantify the irrigation intervals and periods of the two species based on a field test of precision irrigation control. This will provide the theoretical basis for developing water-saving irrigation measures in the shelterbelt.

2. Materials and Methods

2.1. Study Area

This study was carried out in the Taklimakan Desert Highway shelterbelt, which was built from Xiaotang to Minfeng, being 436 km long and 72–78 m wide (Figure 1a). It is characterized by an extremely high temperature, less rainfall, and strong evaporation. According to the Tazhong meteorological station (83°48′14.169″ E,38°54′2.038″ N), the average annual temperature in this area is 12.4 °C. The extreme minimum temperature is −22.2 °C, and the extreme maximum temperature reaches 45.6 °C. The average annual rainfall is 24.6 mm, while the average annual evaporation is 3639 mm. The average relative humidity is only 29.4% (Figure 1b). The average annual wind speed is 2.5 m s^{-1}, and the maximum instantaneous wind speed reaches 20 m s^{-1} [15]. The soil type is mobile aeolian soil, and the soil salt content is 1.26–1.63 g kg^{-1} [19]. The soil physical properties in this area have been described by Zhang et al. [2].

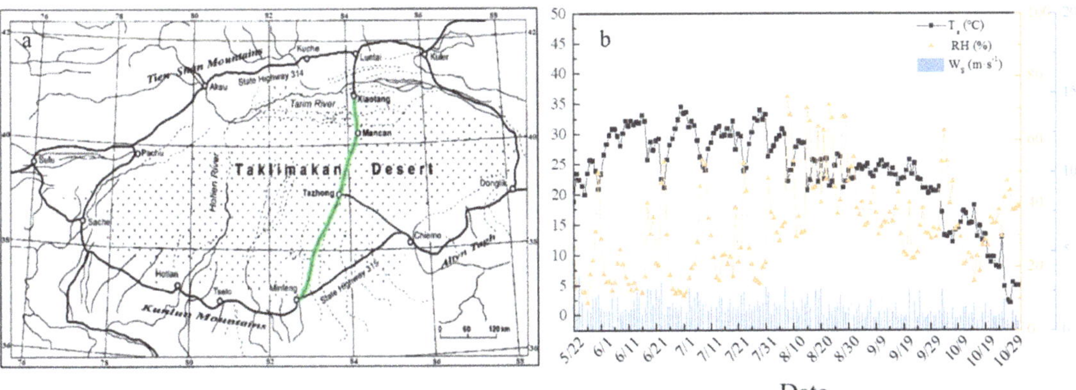

Figure 1. Taklimakan Desert Highway (**a**), and atmospheric temperature (T_a), air humidity (RH), and wind speed (W_S) at the height of 2 m in the middle of the Taklimakan Desert in 2016 (**b**).

The vegetation community is very sparse, and most areas have no vegetation [18]. *Haloxylon ammodendron, Calligonum mongolicum,* and *Tamarix L* are the main three species in the Taklimakan Desert Highway shelterbelt (Figure 2). Saline groundwater was used for drip irrigation. The salinity of irrigation water was 4.03 g L^{-1}; the irrigation period was one time every 10 days in July and August, and 15 days in other months, while there was no irrigation in winter (from November to February of the following year). The irrigation amount was 35 mm each time [11].

Figure 2. Three main species in Taklimakan Desert Highway shelterbelt (adapted from Zhang et al. [2]).

2.2. Experiment Design and Data Processing

This experiment selected six rows of well-growing trees in the Taklimakan Desert Highway shelterbelt for irrigation treatment, with each row of approximately 100 m. *Haloxylon ammodendron* and *Calligonum mongolicum* were planted 8 years ago. The same irrigation amount was adopted, and a small switch was installed on the drip irrigation pipe to control the irrigation period. The distance between two rows of plants was at least 5 m. Two rows of trees were planted in each row with a 1 m spacing. Drip irrigation pipes were laid close to the trunk, and the water outlet holes were spaced by 1 m. Three irrigation levels were set, W1 = 17.5 mm, W2 = 25 mm, and W3 = 35 mm, and the three irrigation periods were F1 = 10 d, F2 = 20 d, F3 = 40 d. The samples were taken from different plants in each treatment plot. Three plants were selected from each treatment plot and one sample was taken from each plant for a total of three repetitions. Moreover, in the experiment design, the W1F1, W1F2, W1F3, W3F1, and W3F2 treatments were selected in order to analyze the effects of different combinations of irrigation amount and irrigation period on soil moisture under the same total amount of irrigation over 40 days. The total irrigation amount of the W1F1 and W3F2 treatments in 40 days was 70 mm. The total irrigation amount of the W1F2 and W3F1 treatments was 35 mm in 40 days, and that of the W1F3 treatment was 17.5 mm in 40 days. The combination of irrigation amount and irrigation period for W1 and W3 only allowed the same combination of total irrigation amount. In contrast, the irrigation level of W2 was different from the total irrigation amount of other treatments within 40 days, so the W2 treatment was not selected in this study. The specific field configuration is shown in Figure 3.

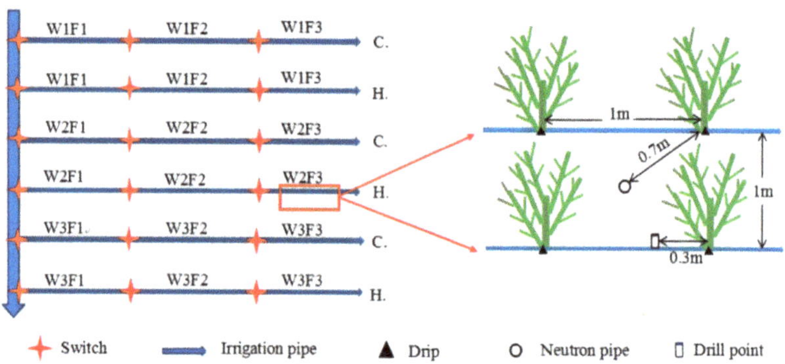

Figure 3. Schematic diagram of field experiment design. (C. means *Calligonum mongolicum*; H. means *Haloxylon ammodendron*; hereinafter the same).

Before the test, three water outlet holes were selected under the drip irrigation pipe in each plot, and a measuring cylinder was placed under each outlet hole to measure the average water output per unit time of each drip irrigation pipe. From 10 August to 20 September 2015, soil samples drilled at 30 cm from the root in each cell were taken at a depth of 0–200 cm (0–10 cm, 10–20 cm, 20–40 cm, 40–60 cm, 60–80 cm, 80–100 cm, 100–120 cm, 120–140 cm, 140–160 cm, 160–180 cm, 180–200 cm) on the 1st, 4th, 9th, 14th, 19th, 29th, and 39th days after irrigation, to evaluate the differences in soil moisture changes under different irrigation strategies with the same irrigation amount of 40 days.

To analyze the change in soil moisture in the whole plot in different months, a 1-m long aluminum tube instrument with LNW-50A neutron probes (CAS, Nanjing, Jiangsu, CHN, 1986) was used for measurement. The neutron tubes were buried in the middle of each plot in 2016. The horizontal distance between the neutron tube and the dropper was 70 cm, and the buried depth was 320 cm. Before the measurement, the neutron instrument was calibrated in layers (0–40 cm and 40–300 cm). It was drilled at a horizontal distance of 30 cm from the dropper to a depth of 200 cm with each 20 cm layer. Each day before irrigation was selected to measure the soil water moisture by drilling and neutron meter in May, July, August, and September of 2016.

After the drilling samples were taken, they were packed into a numbered aluminum box and weighed immediately. Then, the samples were returned to the laboratory and dried in an oven until the constant weight was determined again. The soil mass moisture content was calculated by the calibration curve.

Data statistics and analysis were performed with Excel and SPSS, and plotted with Origin software. Inter-treatment comparisons were compared by one-way ANOVA.

3. Results

3.1. Effect of Irrigation Period on the Spatial Distribution of Soil Water Content

To compare the effects of different irrigation regimes on soil water content, the W1F1, W1F2, W3F1, W3F1, W3F2, and W3F3 treatments in July 2016 (10 d after irrigation) have been selected to analyze the spatial distribution characteristics of soil water content (Table 1).

Table 1. Experimental treatments.

Processing Number	Single Irrigation Volume (mm)	Irrigation Period (d)	Total Irrigation Volume at 40 Days (mm)
W3F1	35	10	140
W3F2	35	20	70
W3F3	35	40	35
W1F1	17.5	10	70
W1F2	17.5	20	35

Under the same irrigation regime, the soil moisture availability shows a clear difference due to the different root growth distributions of *Haloxylon ammodendron* and *Calligonum mongolicum* (Figure 4). In the vertical depth, the soil water content of *Calligonum mongolicum* at 0–100 cm is greater than 100–200 cm in all treatments. The soil water content of *Haloxylon ammodendron* at 0–100 cm in the W3F1 and W3F2 treatments is less than 100–200 cm, and vice versa under other treatments. At the horizontal distance, the soil water content of *Calligonum mongolicum* at 30 cm under the W3F1 and W3F2 treatments is less than 70 cm, and vice versa under other treatments. The soil water content of *Haloxylon ammodendron* at 30 cm under the W3F1, W3F2, and W1F1 treatments is less than 70 cm, and vice versa under other treatments.

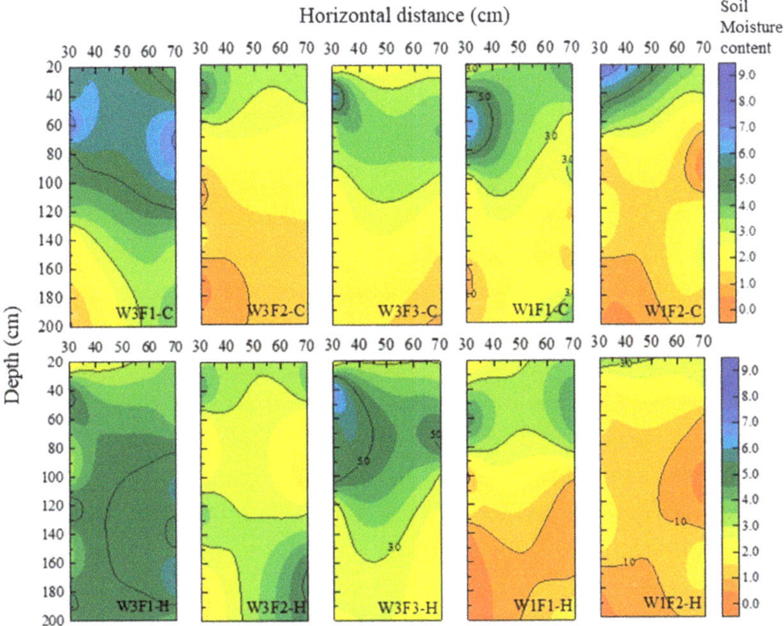

Figure 4. Spatial variation in soil moisture content under different irrigation periods. The soil moisture content in the low (W1), medium (W2), and high (W3) irrigation amounts. F means irrigation frequency with F1 = 10 d, F2 = 20 d, F3 = 40 d. C and H represent *C. mongolicum* and *H. ammodendron*, respectively.

One-way ANOVA is used to compare the differences in soil water content between the treatments shown in Table 2. The soil samples taken at different soil depths obey normal distribution according to the Kolmogorov–Smirnov test. In addition, the F-test of equality of variances with the Least Significance Difference (LSD) test shows the homogeneity of variance. Therefore, one-way ANOVA is suitable for statistical analysis. One-way ANOVA displays the differences in soil water content between treatments at the 0.05 significance level. The mean soil water content of *Calligonum mongolicum* at 0–200 cm is W3F1 > W1F1 > W3F3 > W1F2 > W3F2, while the mean soil water content of *Haloxylon ammodendron* at 0–200 cm is W3F1 > W3F3 > W3F2 > W1F1 > W1F2. Under a 35 mm single irrigation amount, the soil moisture content varies significantly between F1 and F2, F1 and F3, but not significantly between F2 and F3 under a 17.5 mm (strain grade) single irrigation amount; the soil water content is F1 > F2, but this is not significant during the study periods. For the same total water irrigation amount, the difference in soil water content between W3F2 and W1F1, W3F3 and W1F2 are insignificant in the *Calligonum mongolicum*, but both are significant in the *Haloxylon ammodendron*. Under a 70 mm total water irrigation amount, the difference in soil water content of *Haloxylon ammodendron* is mainly focused at the 100–200 cm layer. Under 35 mm water irrigation, the soil moisture of the two species is significantly different between 0–100 cm and 100–200 cm.

Table 2. Soil moisture content between different treatments ($p < 0.05$).

Treatment	Soil Water Content %					
	0–100 cm		100–200 cm		0–200 cm	
	Average Value	Standard Deviation	Average Value	Standard Deviation	Average Value	Standard Deviation
W3F1-C	5.73 a	1.08	2.93 a	1.29	4.33 a	1.84
W3F2-C	2.72 b	1.10	1.15 b	0.46	1.93 b	1.15
W3F3-C	3.37 b	1.26	1.70 b	0.45	2.53 b	1.26
W1F1-C	3.92 b	1.41	1.93 b	1.01	2.93 b	1.58
W1F2-C	2.95 b	2.45	1.57 b	0.81	2.26 b	1.95
W3F1-H	3.97 ab	1.03	4.90 a	0.70	4.44 a	0.99
W3F2-H	2.69 bc	0.91	3.43 b	1.17	3.06 b	1.11
W3F3-H	4.43 a	1.42	2.43 c	0.34	3.43 b	1.44
W1F1-H	3.09 ab	1.32	0.86 d	0.71	1.97 c	1.54
W1F2-H	1.76 c	1.01	1.20 d	0.59	1.48 c	0.87

Note: A lowercase letter indicates a significant difference ($p = 0.05$) after irrigation at the different intervals. The markers (a, b, c) with different letters differ significantly ($p < 0.05$) according to Least Significance Difference test.

3.2. Temporal Variation in Soil Moisture Profile during Irrigation Period under Different Irrigation Regimes

According to the W1F1, W1F2, W3F1, W3F2, and W3F3 treatments, the dynamic changes in soil water content during the irrigation period are analyzed. The soil surface water content reaches the maximum on the first day after irrigation. The water moisture at the 0–60 cm layer gradually decreases with the temporal variation after irrigation. Subsequently, the variation range of soil moisture decreases with the increase in soil depth. The soil moisture above 60 cm is greatly affected by meteorological factors, while the soil moisture below 60 cm is mainly influenced by water redistribution and water absorption by roots (Figure 5).

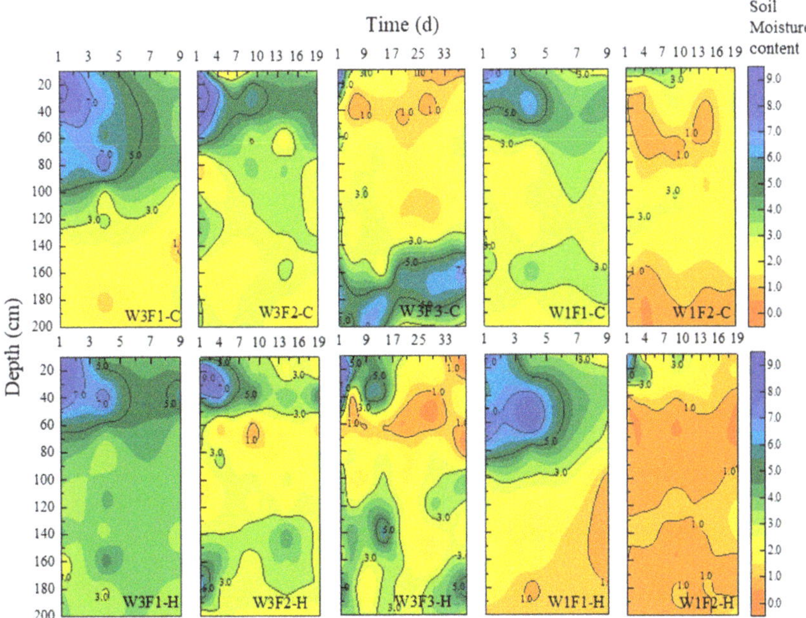

Figure 5. Temporal variation in soil moisture content in different treatments. The soil moisture content in the low (W1), medium (W2), and high (W3) irrigation amounts. F means irrigation frequency with F1 = 10 d, F2 = 20 d, F3 = 40 d. C and H represent *C. mongolicum* and *H. ammodendron*, respectively.

Therefore, the soil is divided into two layers for analysis: 0–60 cm (shallow layer) and 60–200 cm (deep layer). The soil water content of 0–60 cm is significantly greater than that of 60–200 cm and decreases rapidly after irrigation, with a decrease rate greater than the soil water content for 60–200 cm. The F1 and F2 irrigation periods decrease rapidly at 1–9 d after irrigation, and the decline rate of deep soil water content is slow or unchanged on the ninth day. The water content in the shallow and deep soil is similar. The treated soil water content under the F2 irrigation period is in a slow decline period at 9–19 d. The soil moisture in the shallow layers under the F3 treatment decreases rapidly during 1–4 d, decreases slowly during 4–9 d, and remains relatively stable in both the shallow and deep layers during 9–39 d (Figure 6).

3.3. Response of Soil Moisture to Irrigation Regime in Different Months

One-way ANOVA is used to compare the soil moisture content at 0–300 cm under different treatments in the same month from the 0–300 cm layer on 22 May, 12 July, 20 August, and 20 September 2016. As shown in Figure 7, the soil water content of *Calligonum mongolicum* and *Haloxylon ammodendron* at 0–300 cm decreases from May to July, and increases from July to September. The lowest water content is observed in July, while the highest is in September.

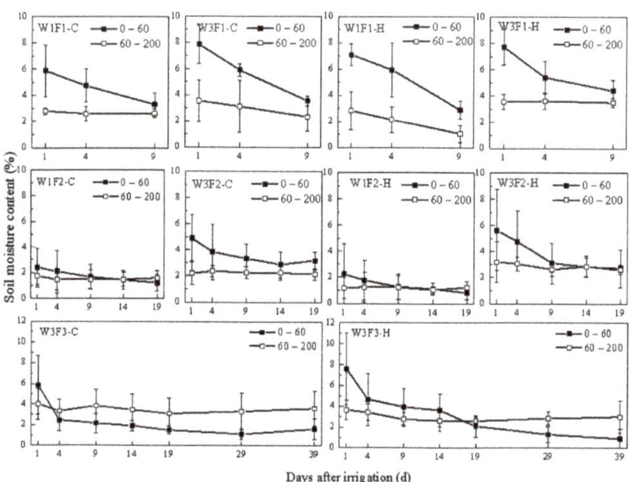

Figure 6. Variation in soil moisture content with the number of days after irrigation.

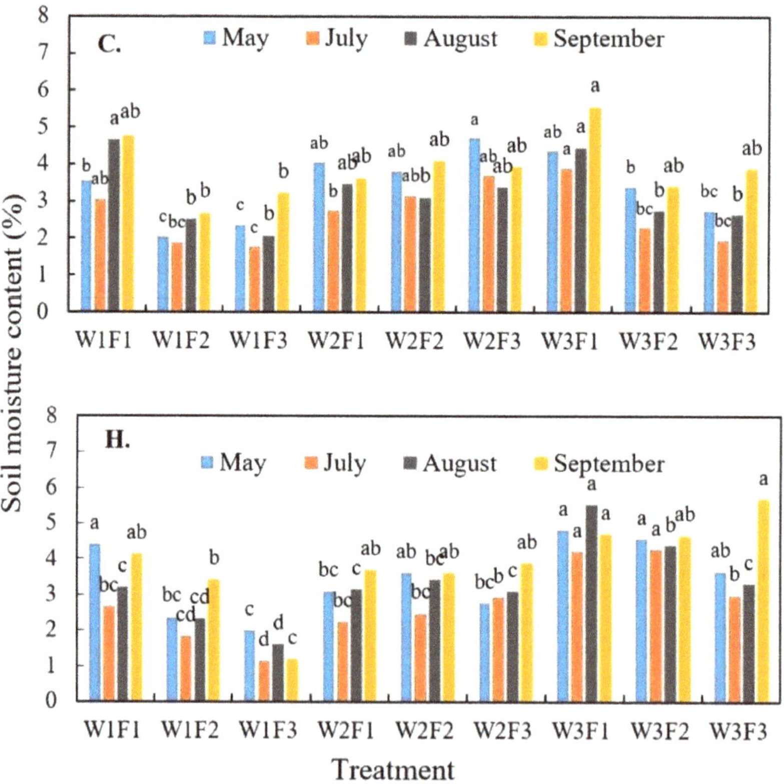

Figure 7. Monthly changes in soil moisture content at 0–300 cm under different treatments. (Lowercase letters indicate differences between treatments in the same month. The markers (a, b, c) with different letters differ significantly ($p < 0.05$) according to Least Significance Difference test).

Table 3 shows the differences in soil water storage between May and September at the 0–300 cm soil layer under each treatment. The water storage of *Calligonum mongolicum* at 0–100 cm increases, except for the W2F3 treatment. The water storage at 100–200 cm increases by 2.26 mm only in the W3F3 treatment, while values for other treatments decrease. The water storage at 200–300 cm increases slightly under the W3F1 and W3F2 treatments, but decreases under other treatments. The difference in water storage at 0–300 cm is F1 > F3 > F2, which are all positive values. Under W2 treatment, the water storage is F2 > F1 > F3, and the water storage is reduced under the W2F1 and W2F3 treatments.

Table 3. Differences in soil water storage between May and September at 0–300 cm soil layer under each treatment (C and H refer to *Calligonum mongolicum* and *Haloxylon ammodendron*, respectively).

Treatment	Difference in Soil Water Storage in September and May (mm)			
	0–100 cm	100–200 cm	200–300 cm	0–300 cm
W1F1-C	51.00	−8.03	−6.62	36.35
W1F2-C	26.71	−3.88	−3.75	19.08
W1F3-C	34.13	−2.22	−5.70	26.21
W2F1-C	7.47	−10.02	−9.44	−11.99
W2F2-C	26.59	−4.44	−13.93	8.22
W2F3-C	−10.95	−3.58	−8.28	−22.81
W3F1-C	34.62	−2.94	4.70	36.37
W3F2-C	4.72	−6.12	1.80	0.40
W3F3-C	31.66	2.62	−0.45	33.83
W1F1-H	−15.54	0.12	6.70	−8.72
W1F2-H	−13.64	23.99	21.04	31.40
W1F3-H	−5.12	−16.35	−2.72	−24.19
W2F1-H	32.22	−11.46	−2.56	18.20
W2F2-H	−8.66	10.34	−1.18	0.50
W2F3-H	36.16	−8.07	5.96	34.05
W3F1-H	14.64	−12.45	8.41	10.60
W3F2-H	4.10	−8.50	7.00	2.60
W3F3-H	15.81	17.32	17.68	50.81

The water storage of *Haloxylon ammodendron* at 0–100 cm decreases under the W1F1, W1F2, W1F3, and W2F2 treatments, while values for the other treatments increase. At 100–200 cm, the water storage volume increases under the W1F1, W1F2, W2F2, and W3F3 treatments. The other treatments' values are reduced. The water storage from 200 to 300 cm is reduced under the W1F3, W2F1, and W2F2 treatments, while values for all other treatments increase. The water storage difference of *Haloxylon ammodendron* at 0–300 cm is F3 > F1 > F2 under W2 > W3 irrigation amount, which are positive values. Under the W1 irrigation amount, the water storage difference is F2 > F1 > F3, and the treatment for W2F3 has a higher value than that for W2F1.

4. Discussion

In arid and semi-arid regions, the water resources directly affect the distribution and growth of plants. In the Taklimakan Desert, the climate is extremely dry, and the annual precipitation (36.6 mm) is far from sufficient to meet the evapotranspiration demand (3638.6 mm). The groundwater depth is more than 10 m, the replenishment effect of groundwater on soil water is negligible, and the main source of soil water is irrigation water [1]. The plants are facing the danger of long-term water shortage. After the Taklimakan Desert Highway shelterbelt was built, the tree species in the shelterbelt were mainly salt-tolerant. The water for plant growth came from underground high-salinity water drip irrigation. At present, although the current drip irrigation system can basically satisfy the growth of tree species in the shelterbelt, the utilization efficiency of plants for irrigation water is low [2].

The study of a reasonable saline irrigation becomes the basis of shelterbelt management and its sustainable existence.

In this study, the water supply rate is specific, and soil water infiltration is determined by soil water infiltration capacity, which is associated with soil wetness and porosity [20,21]. The surface soil moisture content is low from June to early August, resulting in slow transverse water transfer and little change due to intense surface evaporation. The growth of plant roots increases the non-capillary pores, improves the soil water conductivity, and lays a foundation for efficient soil water transport and storage [22]. Due to the downward growth of plant roots and the formation of soil macropores, the irrigation water can quickly reach the deep soil, causing periodic changes in soil water content. The lower limit of soil water evaporation is 40–60 cm, and excessive surface water content increases water evaporation, which is not conducive to water storage. The deeper the soil layer, the lower the soil moisture. Affected by atmospheric evaporation and water absorption by plant roots, the soil water storage variation coefficient is smaller. The results are consistent with the previous studies, such as Li et al. [1] and Zhang et al. [2].

After irrigation, the variation in soil water can be divided into a period of rapid water decline, slow water decline, and a relatively stable water level. Due to the loose soil in the sandy land, the shallow soil water can quickly infiltrate after irrigation. Subsequently, the strong evaporation effect leads to a dry sand layer forming on the soil surface, which significantly inhibits soil moisture evaporation [23]. The water absorption of plants mainly causes a decrease in soil moisture. When the soil moisture is relatively stable, soil moisture at the 0–200 cm layer cannot meet the needs of plant growth, and the water demand of plants mainly comes from the deep soil water supply, while soil moisture maintains a relatively stable state [2].

We noted the largest differences in soil water content between the two plant types in July, followed by August, May, and September. Soil water dissipation mainly includes soil evaporation and plant transpiration. In July, the temperature is the highest, and the water requirement for plant transpiration and soil evaporation is the largest, so the difference between treatments is the largest. In September, the temperature decreases, plant growth slows down, the water requirement decreases, and the difference between treatments is the least.

Under the irrigation regime with a 35 mm irrigation amount, the plants grow well, and part of the soil evaporation is reduced by shading. Irrigation can not only meet the needs of plant growth but also replenish soil water. Under the irrigation regime with a 17.5 mm irrigation amount, the plant growth is weakened due to drought stress in the early stage, although the irrigation amount is small. The leaf transpiration and root water absorption are reduced, and water dissipation is weakened. Moreover, insufficient soil water partially inhibits soil evaporation [24]. Therefore, this study highlights that an irrigation regime with a 35 mm irrigation amount is beneficial to soil water storage. More irrigation water will infiltrate into the deep soil, which will not be absorbed and utilized by plants, resulting in water waste. Increasing the single irrigation amount and prolonging the irrigation period can allow the more effective use of irrigation water. This study highlights that saline groundwater irrigation provides potential advantages for desert plants' survival under reasonable irrigation regimes.

The saving of water and improvement of water use efficiency are undoubtedly fundamental problems associated with such drought regions to avoid lowering the groundwater levels and to prevent ecological degradation. Although desert plants have strong resistance to water and saline stresses and different stress adaptation mechanisms at different growth stages [25], this study evaluated water dynamics and irrigation regimes only under preset irrigation combinations. Further work should conduct additional studies to examine whether or not an appropriate smaller amount of repeated irrigation will increase the water use efficiency of plants and reduce the ineffective evaporation of water. In addition, soil water infiltration and evaporation lead to salinity storage in the soil. The presence and accumulation of salinity affect plants' physiological ecology. The plants' adaptation to saline water irrigation and their responses to the different irrigation regimes should be considered in a future study.

5. Conclusions

Based on a field test of precision irrigation control in the Taklimakan Desert Highway shelterbelt, this study examined the effects of irrigation regimes on the soil water dynamics of two typical woody halophyte species (*Haloxylon* and *Calligonum*), and quantified the irrigation intervals and periods. The effects of saline water irrigation regimes on the soil water dynamics of two typical woody halophyte species (i.e., *Calligonum mongolicum* and *Haloxylon ammodendron*) show that:

(1) Their soil water content at 100–200 cm is significantly greater under the irrigation regime with a 17.5 mm irrigation amount than that under a 35 mm irrigation amount. Increasing the amount of water for single irrigation and prolonging the irrigation period will lead to more effective irrigation water use. After irrigation, the change in soil moisture of the two species in the shallow 0–60 cm layer can be divided into a rapid decline period (1–9 d), a slow decline period (9–19 d), and a relatively stable period (19–39 d). The decrease rate of soil moisture at 0–60 cm depth is significantly higher than that at 60–200 cm depth. The soil moisture below 200 cm replenishes the soil moisture below 60 to 200 cm.

(2) From May to July, the plant growth is vigorous and the temperature gradually increases, and the soil water is in the net consumption stage. The plant growth rate slows down from July to September, and the temperature decreases. The soil water is in the net replenishment stage; the difference in soil water content between 0 and 300 cm is the largest in July and August and the smallest in September.

(3) The irrigation regime combining a 35 mm irrigation amount with 10 days benefits soil water storage and water content with respect to *Calligonum*, while the irrigation regime combining a 35 mm irrigation amount with 40 days is best for *Haloxylon*.

This study highlights that saline groundwater irrigation is advantageous for supporting desert plants' survival and preventing ecological degradation under reasonable irrigation regimes. Future work should focus on the plants' adaptation to saline water irrigation and their responses to the different irrigation regimes and water-saving irrigation measures in the desert shelterbelt construction.

Author Contributions: Conceptualization and methodology, J.L., Y.Z., J.Z. and J.X.; data analysis, J.L. and Q.H.; writing—original draft preparation, J.L.; writing—review and editing, Y.Z., J.Z. and J.X. All authors have read and agreed to the published version of the manuscript.

Funding: This research was supported by the National Natural Science Foundation of China (41977009, 41877541, 41471222, and 42071259), the original innovation project of the basic frontier scientific research program, Chinese Academy of Sciences (ZDBS-LY-DQC031), the Natural Science Foundation of Xinjiang Uygur Autonomous Region (2021D01E01), the Third Batch of Tianshan Talents Program of Xinjiang Uygur Autonomous Region (2021–2023), the Youth Innovation Promotion Association of the Chinese Academy of Sciences (2019430), and the State Key Laboratory of Desert and Oasis Ecology, Xinjiang Institute of Ecology and Geography, Chinese Academy of Sciences (G2018-02-08).

Data Availability Statement: The data are available from the corresponding author upon reasonable request.

Conflicts of Interest: The authors declare no conflict of interest.

References

1. Li, C.; Lei, J.; Zhao, Y.; Xu, X.; Li, S. Effect of saline water irrigation on soil development and plant growth in the Taklimakan Desert Highway shelterbelt. *Soil Tillage Res.* **2015**, *146*, 99–107. [CrossRef]
2. Zhang, J.G.; Lei, J.Q.; Wang, Y.D.; Zhao, Y.; Xu, X.W. Survival and growth of three afforestation species under high saline drip irrigation in the Taklimakan Desert, China. *Ecosphere* **2016**, *7*, e01285. [CrossRef]

3. Zhang, B.; Kang, S.; Li, F.; Zhang, L. Comparison of three evapotranspiration models to Bowen ratio-energy balance method for a vineyard in an arid desert region of northwest China. *Agric. For. Meteorol.* **2008**, *148*, 1629–1640. [CrossRef]
4. Dai, S.Y.; Lei, J.Q.; Zhao, J.F.; Fan, J.; Fan, D.; Zeng, F.; Ye, K. The characteristics of groundwater and their effects on eco-environment of desert-oasis transitional zone in western Qira of China. *J. Arid. Land Resour. Environ.* **2009**, *23*, 4782–4786.
5. Chartzoulakis, K.S. Salinity and olive: Growth, salt tolerance, photosynthesis and yield. *Agric. Water Manag.* **2005**, *78*, 108–121. [CrossRef]
6. Chen, M.; Kang, Y.; Wan, S.; Liu, S. Drip irrigation with saline water for oleic sunflower (*Helianthus annuus* L.). *Agric. Water Manag.* **2009**, *96*, 1766–1772. [CrossRef]
7. Feikema, P.M.; Morris, J.D.; Connell, L.D. The water balance and water sources of a Eucalyptus plantation over shallow saline groundwater. *Plant Soil* **2010**, *332*, 429–449. [CrossRef]
8. Xu, X.; Li, B.; Wang, X. Progress in study on irrigation practice with saline groundwater on sandlands of Taklimakan Desert hinterland. *Chinese Sci. Bull.* **2006**, *51*, 161–166. [CrossRef]
9. Tedeschi, A.; Dell'Aquila, R. Effects of irrigation with saline waters, at different concentrations, on soil physical and chemical characteristics. *Agric. Water Manag.* **2005**, *77*, 308–322. [CrossRef]
10. Huang, C.H.; Xue, X.; Wang, T.; De Mascellis, R.; Mele, G.; You, Q.G.; Peng, F.; Tedeschi, A. Effects of saline water irrigation on soil properties in northwest China. *Environ. Earth Sci.* **2010**, *63*, 701–708. [CrossRef]
11. Zhang, J.; Xu, X.; Lei, J.; Sun, S.; Fan, J.; Li, S.; Gu, F.; Qiu, Y.; Xu, B. The salt accumulation at the shifting aeolian sandy soil surface with high salinity groundwater drip irrigation in the hinterland of the Taklimakan Desert. *Sci. Bull.* **2008**, *53*, 63–70. [CrossRef]
12. Wang, X.; Xu, X.; Lei, J.; Li, S.; Wang, Y. The vertical distribution of the root system of the desert highway shelterbelt in the hinterland of the Taklimakan Desert. *Sci. Bull.* **2008**, *53*, 79–83. [CrossRef]
13. Li, S.; Lei, J.; Xu, X.; Wang, H.; Fan, J.; Gu, F.; Qiu, Y.; Xu, B.; Gong, Q.; Zheng, W. Topographical changes of ground surface affected by the shelterbelt along the Tarim Desert Highway. *Sci. Bull.* **2008**, *53*, 8–21. [CrossRef]
14. Li, C.; Shi, X.; Mohamad, O.A.; Gao, J.; Xu, X.; Xie, Y. Moderate irrigation intervals facilitate establishment of two desert shrubs in the Taklimakan Desert Highway Shelterbelt in China. *PLoS ONE* **2017**, *12*, e0180875. [CrossRef] [PubMed]
15. Ding, X.; Zhou, Z.; Xu, X.; Wang, Y. Dynamics of Soil Water and Salt in Soil under Artificial Plantation Shelterbelt Drip-irrigated with Saline Water in the Center of the Taklimakan Desert. *Acta Pedol. Sin.* **2016**, *53*, 103–116.
16. Fu, S.; Hu, S.; Hao, L.; Wang, Z. Water sources of dominant plants in haloxylon ammodendron community at the southern edge of gurbantunggut desert. *J. Desert Res.* **2018**, *38*, 1024–1032.
17. Zhang, Y.M.; Pan, H.X.; Pan, B.R. Distribution Characteristics of Biological Crust on Sand Dune Surface in Gurbantunggut Desert, Xinjiang. *J. Soil Water Conserv.* **2004**, *18*, 61–64.
18. Beltrán, J.M. Irrigation with saline water: Benefits and environmental impact. *Agric. Water Manag.* **1999**, *40*, 183–194. [CrossRef]
19. Verma, A.; Gupta, S.; Isaac, R. Use of saline water for irrigation in monsoon climate and deep water table regions: Simulation modeling with SWAP. *Agric. Water Manag.* **2012**, *115*, 186–193. [CrossRef]
20. Han, W.; Cao, L.; Yimit, H.; Xu, X.; Zhang, J. Optimization of the saline groundwater irrigation system along the Tarim Desert Highway Ecological Shelterbelt Project in China. *Ecol. Eng.* **2012**, *40*, 108–112. [CrossRef]
21. Guo, Z. Estimating Method of Maximum Infiltration Depth and Soil Water Supply. *Sci. Rep.* **2020**, *10*, 9726. [CrossRef] [PubMed]
22. Jiang, X.J.; Liu, W.; Chen, C.; Liu, J.; Yuan, Z.-Q.; Jin, B.; Yu, X. Effects of three morphometric features of roots on soil water flow behavior in three sites in China. *Geoderma* **2018**, *320*, 161–171. [CrossRef]
23. Ma, X.D.; Chen, Y.N.; Zhu, C.G.; Li, W.H. The variation in soil moisture and the appropriate groundwater table for desert riparian forest along the Lower Tarim River. *J. Geogr. Sci.* **2011**, *21*, 150–162. [CrossRef]
24. Duan, Z.H. Study on Soil Water Cycle Under Irrigation in Takelamagan Desert. *J. Arid. Land Resour. Environ.* **2001**, *15*, 63–67.
25. Resco, V.; Ignace, D.D.; Sun, W.; Huxman, T.E.; Williams, J.F.W.G. Chlorophyll fluorescence, predawn water potential and photosynthesis in precipitation pulse-driven ecosystems: Implic. *Ecol. Studies. Funct. Ecol.* **2008**, *22*, 479–483. [CrossRef]

Article

Effect of Shelterbelt Construction on Soil Water Characteristic Curves in an Extreme Arid Shifting Desert

Chuanyu Ma [1], Luobin Tang [1], Wenqian Chang [1], Muhammad Tauseef Jaffar [1], Jianguo Zhang [1,2,*], Xiong Li [1], Qing Chang [2] and Jinglong Fan [2]

1 Key Laboratory of Plant Nutrition and the Agri-Environment in Northwest China, Ministry of Agriculture, Northwest A&F University, Xianyang 712100, China; chuanyuma1998@163.com (C.M.); tangluobin2020@163.com (L.T.); changwenqian1991@163.com (W.C.); tauseefjaffar8555@gmail.com (M.T.J.); zhengfadeshui@163.com (X.L.)

2 Taklimakan Desert Research Station, Xinjiang Institute of Ecology and Geography, Chinese Academy of Sciences, Korla 841000, China; changqing@ms.xjb.ac.cn (Q.C.); fanjl@ms.xjb.ac.cn (J.F.)

* Correspondence: zhangjianguo21@nwafu.edu.cn

Abstract: To explore the impact of artificial shelterbelt construction with saline irrigation on the soil water characteristic curve (SWCC) of shifting sandy soil in extreme arid desert areas, three treatments including under the shelterbelt (US), bare land in the shelterbelt (BL) and shifting sandy land (CK) in the hinterland of the Taklimakan Desert were selected. The age of the shelterbelt is 16, and the vegetation cover is mainly *Calligonum mongolicum*. The soils from different depths of 0–30 cm were taken keeping in view the objective of the study. The SWCCs were determined by the centrifugal method and fitting was performed using various models such as the Gardner (G) model, Brooks–Corey (BC) model and Van Genuchten (VG) model. Then, the most suitable SWCC model was selected. The results showed that electrical conductivity (EC) and organic matter content of BL and US decreased with the increasing soil depth, while the EC and organic matter content of CK increased with the soil depth. The changes in soil bulk density, EC and organic matter of 0–5 cm soil were mostly significant ($p < 0.05$) for different treatments, and the differences in SWCCs were also significant among different treatments. Moreover, the construction of an artificial shelterbelt improved soil water-holding capacity and had the most significant impacts on the surface soil. The increase in soil water-holding capacity decreased with increasing soil depth, and the available soil water existed in the form of readily available water. The BC model and VG model were found to be better than the G model in fitting results, and the BC model had the best fitting result on CK, while the VG Model had the best fitting result on BL with higher organic matter and salt contents. Comparing the fitting results of the three models, we concluded that although the fitting accuracy of the VG model tended to decrease with increasing organic matter and salinity, the VG model had the highest fitting accuracy when comparing with BC and G models for the BL treatment with high organic matter and salinity. Therefore, the influence of organic matter and salinity should be considered when establishing soil water transfer function.

Keywords: aeolian sandy soil; physiochemical properties; soil water-holding capacity; model fitting; Taklimakan Desert

Citation: Ma, C.; Tang, L.; Chang, W.; Jaffar, M.T.; Zhang, J.; Li, X.; Chang, Q.; Fan, J. Effect of Shelterbelt Construction on Soil Water Characteristic Curves in an Extreme Arid Shifting Desert. *Water* **2022**, *14*, 1803. https://doi.org/10.3390/w14111803

Academic Editor: Ognjen Bonacci

Received: 27 April 2022
Accepted: 31 May 2022
Published: 2 June 2022

1. Introduction

Soil moisture is an important factor affecting plant growth and is a major driving force for the sustainability of many terrestrial ecosystems. Moisture changes have significant impacts on vegetation and soil properties [1–3]. Especially in arid and semi-arid regions, soil moisture availability is one of main factors limiting the type and quantity of vegetation, and the water deficiency can lead to severe degradation of vegetation and reduce vegetation cover [4,5]. Therefore, it is important to understand the soil moisture change pattern for the maintenance of vegetation in arid desert areas [6].

The Taklamakan Desert is located in the hinterland of the Eurasian continent in Chinese southern Xinjiang, and is the largest desert in China and the second largest shifting desert in the world. Southern Xinjiang is one of the poorest regions with an extreme drought climate in China. To accelerate the development of regional social economy, the Taklimakan Desert Highway (TDH) was completed in 1995. TDH is 522 km long running across the Taklimakan Desert from north to south and is the longest highway crossing a moving desert in the world. However, serious sand disasters are a great threat to TDH. Thus, in 2003, a shelter–forest belt (the Taklimakan Desert Highway Shelterbelt (TDHS)) having a length of 436 km was constructed on both sides along the highway, which was dominated by shrubs to protect the highway from shifting sand [7]. Low-quality saline groundwater is used for irrigation in order to ensure the survival of shelterbelt plants [8–11]. The regional ecological environment on both sides of the highway has been greatly improved by the fixation of the shifting sand dunes with the artificial shelterbelt [12]. However, saline groundwater irrigation may aggravate soil salinization in future and is harmful for shelterbelt plants [13,14]. Therefore, it is necessary to study the water and salt transport of shelterbelt soils which is helpful for the sustainable utilization and management of TDHS [15].

SWCC, as a component of the unsaturated soil mechanics framework, provides the information needed to characterize the properties of unsaturated soils [16], is an interpretation of the basic constitutive relationships of unsaturated soil phenomena [17], and is an important tool for studying the properties of unsaturated soils [18]. SWCC is one of basic hydraulic properties that simulate water and solute transport under unsaturated conditions [19]; is universally used in agriculture [20], soil physics [21], soil chemistry [22], mineralogy research [23], geotechnical engineering [24]; and is widely used in the soil–plant–atmosphere continuum (SPC) [18,25] and other fields. Due to its importance in soil hydrodynamics and solute transport modeling, many SWCC models, both numerical and theoretical, have been developed [21]. A good SWCC model should have simple and clear parameters and be easy to use. It can satisfy the three characteristics of accuracy, universality and simplicity as much as possible at the same time [26]. The VG model [27], BC model [28] and G model [29] have relatively few parameters, and can accurately describe the SWCC of various soil textures [30–38]. SWCC is influenced by soil texture, bulk density, organic matter, salinity, temperature, etc. [19,30–33,39–42]. Therefore, the fitting results of the models are often different in various study areas. For example, some scholars [43,44] pointed out that the G model can accurately fit SWCC; however, others [45] came to the opposite conclusion, pointing out that the G model cannot accurately fit SWCC. Matlan et al. [26] compared four models and found that the BC model has the most accurate description of the SWCC of sandy soils. Li et al. [46] also proved this, pointing out that the BC model is more suitable than the VG model on soils with high sand content and low clay content. For the SWCC of aeolian sand covered by biocrust, the fitting effect of the VG model is better than the BC model [47].

The literature on SWCC in arid and semi-arid regions is relatively limited as compared to farmland and forest ecosystems. The influence of artificial shelterbelts and long-term saline irrigation on SWCC of sandy soils and the applicability of SWCC models is still unclear. This hinders our understanding towards the soil water-holding capacity and water availability of shelter forests. In this study, we firstly assumed that artificial shelterbelts and long-term saline irrigation have impacts on the water-holding capacity and water availability of different soil layers. In addition, we assumed that the VG model, BC model, and G model have different accuracies in fitting SWCC. Therefore, our study collected soils from 0 to 30 cm layers under the shelterbelt (US), bare land in the shelterbelt (BL) and shifting sandy land (CK) in the hinterland of the Taklimakan Desert, and their SWCCs were determined by the centrifugal method. Combined with its bulk density, organic matter, salinity and other properties, SWCC, pore distribution and soil moisture were analyzed, and the VG model, BC model and G model were used and compared to fit the SWCCs. We aim to reveal the impact of artificial shelterbelt construction on SWCC on shifting sandy soil in extreme arid deserts under saline drip irrigation, so as to provide a basis for desert shelterbelt construction and sustainable management.

2. Materials and Methods

2.1. Study Area

The sampling area is located at the Taklamakan Desert Research Station of the Chinese Academy of Sciences in the hinterland of Taklamakan Desert (39°01′ N, 83°36′ E, 1100 m.a.s.l.), Xinjiang, China (Figure 1). The study area belongs to a warm temperate arid climate. The annual average temperature is 12.4 °C. December is the coldest month with an average monthly temperature of −8.1 °C and July is the hottest month with an average monthly temperature of 28.2 °C. Annual precipitation is 24.6 mm, average relative humidity is 29.4% and annual potential evaporation is up to 3638.6 mm. Annual average wind speed is 2.5 m·s^{-1}, and the maximum instantaneous wind speed is up to 20 m·s^{-1}. The soil type is Xeric Quartzipsamments [48] (Soil Survey Staff 2014), derived from shifting eolian sand, and the basic properties are shown in Table 1.

Figure 1. Location of the study area.

Table 1. Basic physiochemical properties of the collected soil samples.

Soil Depth (cm)	Location	Bulk Density (g/cm^3)	EC (μS/cm)	pH	Organic Carbon (g/kg)	Soil Mechanical Composition (%)			Soil Texture
						Sand (0.05–2 mm)	Silt (0.002–0.05 mm)	Clay (0–0.002 mm)	
0–5	CK	1.55 a	451 c	9.39 a	0.208 c	89.05	2.83	8.12	Sandy
	BL	1.43 b	19,215 a	8.73 a	1.765 b	82.38	9.01	8.61	Loamy sand
	US	1.25 c	10,175 b	9.15 a	3.222 a	65.33	25.46	9.21	Sandy loam
5–10	CK	1.49 b	762 c	8.69 a	0.379 a	89.34	2.73	7.93	Sandy
	BL	1.53 a	4340 b	8.8 a	1.15 a	86.92	4.81	8.27	Loamy sand
	US	1.44 c	6680 a	8.95 a	1.715 a	87.94	4.36	7.7	Loamy sand
10–20	CK	1.47c	2455 a	8.62 b	0.788 a	90.17	1.84	7.99	Sandy
	BL	1.51a	854.5 c	8.71 b	1.058 a	86.09	4.63	9.28	Loamy sand
	US	1.5b	1157 b	9.6 a	1.182 a	88.48	3.34	8.18	Loamy sand
20–30	CK	1.47 b	3240 a	8.82 c	0.83 ab	87.97	4.87	7.16	Sandy
	BL	1.6 a	265 c	9.52 b	0.412 b	87.49	2.7	9.81	Loamy sand
	US	1.44 c	951.5 b	10.11 a	0.912 a	88.09	1.31	10.6	Loamy sand

Note: CK: shifting sandy land; BL: bare land without vegetation cover in the shelterbelt; US: under the shelterbelt. Different lowercase letters represent the significant differences between different treatments ($p < 0.05$). (Soil texture was classified according to USDA standards based on actual measured soil mechanical composition).

2.2. Sample Collection and Determination

In August 2020, undisturbed soil samples were collected with a cutting ring of 100 cm^3 from the soil layers of 0–5 cm, 5–10 cm, 10–20 cm and 20–30 cm under US, BL and CK, and all samples had three replicates (the pictures of the sampling points are shown in Figure 2). The ring knife (with soil) samples were soaked in deionized water for 24 h until saturation, weighed, and then the SWCCs were determined using a high-speed centrifuge (CR 22G III model, Hitachi, Japan). Suction values were determined in lab using a centrifuge with different speeds and time settings. The selected suction values were 10.2 cm (310 r/min, 10 min), 30.6 cm (540 r/min, 12 min), 51 cm (690 r/min, 17 min), 71.4 cm (820 r/min, 21 min), 102 cm (980 r/min, 26 min), 204 cm (1200 r/min, 28 min), 612 cm (2190 r/min, 49 min), 1020 cm (3100 r/min, 57 min), 4080 cm (6200 r/min, 77 min) and 8160 cm (8770 r/min, 87 min), respectively. The weight of the ring knife sample was weighed after the completion of centrifugation at each suction value. After centrifugation, the ring knife was dried in an oven (105 °C) and then weighed, and the water content at the corresponding suction value was obtained from the difference in weights. The moisture under different suction values was plotted according to the SWCC. Meanwhile, soil samples of corresponding layers were also collected with a soil drill and brought to the laboratory and air dried. Part of the samples was passed through a 2 mm sieve to determine soil pH, EC (with an EC500 pH/Conductivity Meter (ExStik, Boston, MA, USA)), and soil mechanical composition was determined using the hydrometer method and soil texture was classified according to USDA standards. Soil particles were divided into sand particles of 2–0.05 mm, silt particles of 0.05–0.002 mm and clay particles of $\leq$0.002 mm, and the left soils were used to measure soil organic matter content with potassium dichromate external heating method after passing through a 0.25 mm sieve. The physical and chemical properties of the soil samples are shown in Table 1.

Figure 2. Photos of sampling points.

2.3. Calculations of Soil Equivalent Pore Sizes and Moisture Constants

The equivalent pores are divided into six levels based on their diameters, including narrow micropores ($\leq$0.3 μm), micropores (0.3–5 μm), fine pores (5–30 μm), medium pores (30–75 μm), macropores (75–100 μm) and interstices ($\geq$100 μm) [47,49].

Based on the measured and fitted SWCC parameters, saturated water θ_s, field capacity θ_f, wilting coefficient θ_r, available water content θ_a and readily available water content θ_{ra} were obtained. While θ_s is the soil water content when water suction is zero, θ_f is the soil water content of when pF = 1.8 (pF is expressed as the logarithm of the centimeter height of the water column of the soil water potential), and θ_r is the soil water content when pF = 4.2, pF = 3.8 represents temporary wilting coefficient. The water content in the range of pF 1.8–4.2 is θ_a, and the water content in the range of pF 1.8–3.8 is θ_{ra} [50].

2.4. SWCC Modeling

VG model, G model and BC model were used to describe the SWCC.

VG model:

$$\theta_{(h)} = \theta_r + \frac{\theta_s - \theta_r}{\left(1 + |\alpha h|^n\right)^m} \tag{1}$$

where $\theta_{(h)}$ is the volume water content, $cm^3\ cm^{-3}$; θ_s is the saturated water content, $cm^3\ cm^{-3}$; θ_r is the residual water content, $cm^3\ cm^{-3}$; h is the matric suction, cm; α is a scale parameter that is related to the inverse of the air entry suction, cm^{-1}; m, n are the fitting parameters; $m = 1 - 1/n$ ($n > 1$).

G model:

$$\theta = \alpha h^{-b} \tag{2}$$

where θ is the volume water content, $cm^3\ cm^{-3}$; h is the matric suction, cm; a and b are fitting parameters.

BC model:

$$S_e = \frac{\theta - \theta_r}{\theta_s - \theta_r} = \begin{cases} (\alpha h)^{-\lambda} & \alpha h > 1 \\ 1 & \alpha h \le 1 \end{cases} \tag{3}$$

where S_e is the effective saturation; λ is the curve shape parameter; the physical meanings of other parameters are the same as above.

2.5. SWCC Model Fitting Accuray Assesments

The coefficient of determination (R^2), root mean square error ($RSME$) and relative error (RE) were used to quantitatively evaluate the fitting effect of the models. Grey correlation analysis allows ranking the importance between different influencing factors [51,52]. The soil physiochemical properties and model parameters were quantitatively analyzed by grey correlational method, and the correlational degree was sorted. The calculation formulas are as follows:

$$R^2 = \frac{\sum_{i=1}^{N} (\theta_i - \overline{\theta_i})^2}{\sum_{i=1}^{N} (\beta_i - \overline{\theta_i})^2} \tag{4}$$

$$RMSE = \sqrt{\frac{\sum_{i=1}^{N} (\theta_i - \beta_i)^2}{N}} \tag{5}$$

$$RE = \frac{|\theta_i - \beta_i|}{\theta_i} \times 100 \tag{6}$$

$$\gamma_{0i} = \frac{1}{n} \sum_{k=1}^{n} \frac{m + \rho M}{\Delta_i(k) + \rho M} \tag{7}$$

In Equations (4)–(6), N is the total number of samples of matric suction; θ_i represents the measured value of soil moisture corresponding to the ith pressure value; $\overline{\theta_i}$ represents the average value of measured soil moisture; β_i is the fitted value of soil moisture corresponding to the ith pressure value. In Equation (7), γ_{0i} is the correlation degree, $\Delta_i(k)$ is the difference sequence, M is the maximum difference sequence, m is the minimum difference sequence, and ρ is the resolution coefficient, which is generally 0.5 in the models [51,52].

RETC software was used to solve and fit the parameters, Excel2010 and Origin2018 were used for data processing and mapping and SPSS18.0 was used for one-way ANOVA and multiple comparisons (a = 0.05, LSD).

3. Results

3.1. Effects of Artificial Shelterbelt Construction on Soil Physiochemical Properties

Soil physiochemical properties were improved after artificial shelterbelt construction. From Table 1, according to USDA system, it is clearly stated that all soil layers of CK are sandy soil, all soil layers of BL treatment are loamy sandy soil, 0–5 cm soil of US treatment was sandy loam soil and 5–30 cm soil was loamy sandy soil. There were significant differences in bulk density and EC among CK, BL and US in the same soil layer ($p < 0.05$). Except for CK, the soil bulk density under BL and US treatments increased with the increasing soil depth, and the soil bulk density of 0–5 cm under BL and US treatments was significantly lower than CK; the bulk density of the 5–30 cm soil layer was the highest in BL, except for 10–20 cm; the bulk density of US treatment was lower than that of CK and BL. The bulk density of the 0–5 cm soil layer in US treatment was the smallest (1.25 g/cm^3), and the highest bulk density at all was 1.6 g/cm^3. With the increase in soil depth, the EC of CK gradually increased, while EC of BL and US treatments decreased with the increasing depth. The difference in EC between BL, US and CK in the 0–5 cm soil layer was the largest, which was 18,764 µS/cm and 9724 µS/cm, respectively. The EC of CK at 10–30 cm was higher than that of BL and US.

3.2. Screening of Soil Water Characteristic Curve Models

As listed in Table 2, R^2 of the fitting values of VG, BC, and G models was ranged between 0.884 and 0.998, which showed the correlations of three models with the measured data were high. As shown in Figure 3, the fitted results of the G model were all higher than the measured points, and the fitted values of the VG model and BC model for BL in the 5–20 cm soil layer were smaller than the measured values.

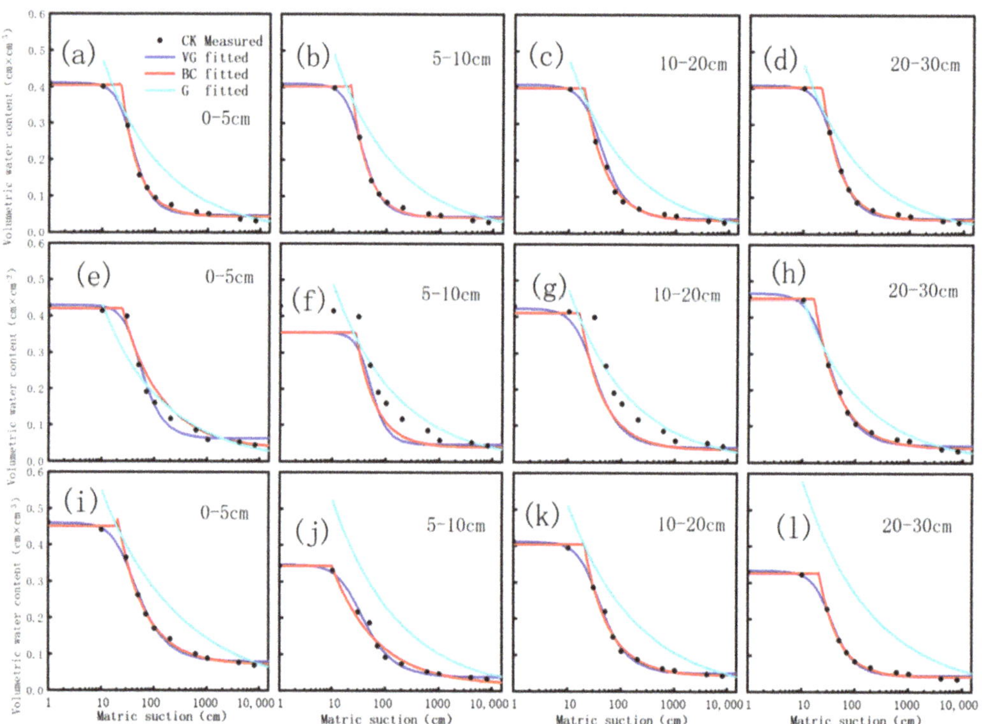

Figure 3. Fitting curves of soil water characteristics of VG model, BC model and G model. (**a–l**) is the serial number of the figure, (**a–d**) is the model fitting result of CK processing 0–5 cm, 5–10 cm, 10–20 cm, 20–30 cm soil layer; (**e–h**) is BL processing 0–5 cm, 5–10 cm, 10–20 cm and the model fitting results of the 20–30 cm soil layer; (**i–l**) is the model fitting result of the US treatment of 0–5 cm, 5–10 cm, 10–20 cm and 20–30 cm soil layers.

The fitting effect of the VG model and BC model for different soil layers of various treatments were always better than that of the G model. The fitting effect of the VG model and BC model was similar for the 10–30 cm soil layers of each treatment. The fitting effect of the BC model for the 0–10 cm soil layer of CK was better than the VG model, while the results of BL were opposite. For US, the BC model had a good fitting effect for the 0–5 cm soil layer, but the fitting effect was contradictory for the 5–10 cm soil layer. For all treatments, the fitting errors of three models increased with the increasing water potential and tended to be stable (Figure 4). The relative errors of the VG model and BC model for different treatments and soil layers were always lower as compared with relative error of the G model.

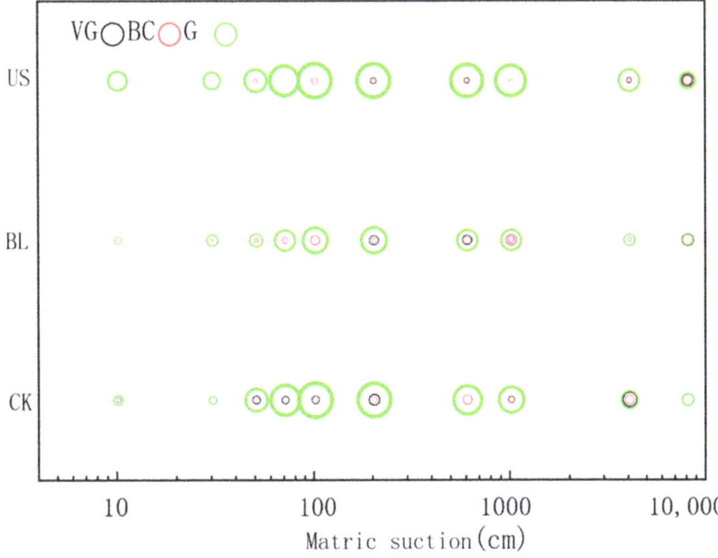

Figure 4. Relative errors of SWCC models under CK, BL and US treatments. (The smaller the circle, the smaller the relative error).

Table 2. R^2 and *RMSE* of different fitting models.

Location	Soil Depth (cm)	VG Model		BC Model		G Model	
		R^2	*RMSE* (%)	R^2	*RMSE* (%)	R^2	*RMSE* (%)
CK	0–5	0.994	0.0102	0.997	0.0079	0.919	0.0682
	5–10	0.996	0.0082	0.998	0.0071	0.906	0.0862
	10–20	0.998	0.0150	0.996	0.0090	0.917	0.0710
	20–30	0.998	0.0631	0.998	0.0059	0.908	0.0636
BL	0–5	0.987	0.0165	0.977	0.0259	0.944	0.0395
	5–10	0.997	0.0067	0.992	0.0117	0.884	0.0787
	10–20	0.997	0.0078	0.998	0.0054	0.935	0.0704
	20–30	0.995	0.0100	0.997	0.0082	0.948	0.0431
US	0–5	0.996	0.0086	0.998	0.0050	0.934	0.0765
	5–10	0.993	0.0089	0.984	0.0147	0.934	0.1133
	10–20	0.997	0.0073	0.995	0.0095	0.913	0.0822
	20–30	0.997	0.0060	0.998	0.0048	0.908	0.1576
	mean	0.9955	0.0140	0.9941	0.0096	0.9206	0.0792

In addition, R^2 and *RMSE* of the VG and BC models showed that the simulating results of CK were better than those of BL and US (Table 2). The results indicated that the increase in salinity and organic matter content may affect the fitness of each model. Table 3 lists the parameters of the three models in SWCC modeling. From grey correlation calculation (Table 4), it could be concluded that VG model parameters a and n have a higher degree of correlation with the soil physiochemical parameters (BC and G models were not included, because the VG model had the best fitting effect). The grey relational degrees between pH, EC, organic matter, bulk density, sand content, silt content, clay content and model parameters (a and n) were all larger than 0.6, and the order of correlation showed as: bulk density > sand content > pH > clay content > organic matter > silt content > EC.

Table 3. Parameters in the modeling of soil water characteristic curves.

Location	Soil Depth (cm)	VG Model				BC Model				G Model		
		θ_r (%)	θ_s (%)	a (cm^{-1})	n	θ_r (%)	θ_s (%)	a (cm^{-1})	λ	a	b	$a \times b$
CK	0–5	0.048	0.411	0.032	2.841	0.044	0.405	0.044	0.728	0.375	0.728	0.273
	5–10	0.044	0.409	0.036	2.805	0.041	0.402	0.047	0.663	0.373	0.663	0.247
	10–20	0.041	0.409	0.033	2.428	0.036	0.400	0.053	0.708	0.379	0.708	0.268
	20–30	0.042	0.409	0.033	2.690	0.039	0.403	0.046	0.733	0.383	0.733	0.281
BL	0–5	0.063	0.430	0.022	2.670	0.043	0.421	0.040	0.990	0.366	0.990	0.362
	5–10	0.047	0.355	0.022	3.291	0.040	0.355	0.037	0.710	0.363	0.710	0.258
	10–20	0.041	0.423	0.044	2.302	0.036	0.409	0.062	0.711	0.379	0.711	0.269
	20–30	0.047	0.467	0.044	2.257	0.040	0.452	0.062	0.826	0.381	0.826	0.315
US	0–5	0.080	0.461	0.031	2.160	0.073	0.451	0.045	0.774	0.290	0.774	0.224
	5–10	0.038	0.347	0.040	2.074	0.016	0.343	0.101	0.616	0.352	0.616	0.217
	10–20	0.051	0.411	0.034	2.295	0.046	0.403	0.049	0.711	0.339	0.711	0.241
	20–30	0.045	0.333	0.035	2.552	0.043	0.326	0.047	0.521	0.328	0.521	0.171

Note: θ_s: saturated water content; θ_r: wilting coefficient; α is a scale parameter that is related to the inverse of the air entry suction; n is the fitting parameter; λ is the curve shape parameter; a and b are fitting parameters; $a \times b$ is the specific water capacity when the soil water suction is 1 Bar.

Table 4. Correlation degree analysis between VG model parameters and soil basic physiochemical parameters.

Correlation Degree	pH	EC (μS/cm)	Organic Carbon (g/kg)	Bulk Density (g/cm^3)	Sand (0.05–2 mm)	Slit (0.002–0.05 mm)	Clay (0–0.002 mm)
a	0.925	0.684	0.783	0.930	0.926	0.744	0.917
n	0.939	0.694	0.794	0.952	0.948	0.781	0.922

3.3. *Effects of Artificial Shelterbelt Construction on Soil Water Retention Performance*

Soil porosity was significantly changed after artificial shelterbelt construction. As listed in Figure 5, few micropores were found under all treatments. Compared with CK, the contents of interstices and macropores in the 0–10 cm soil layer under BL and US were much lower, and the contents of fine pores and micropores were higher. The medium pores of the BL increased the most, with 0–5 cm increased by 3.74% and 5–10 cm increased by 4.94%. Fine pores of US increased the most: 0–5 cm increased by 4.54% and 5–10 cm increased by 2.32%.

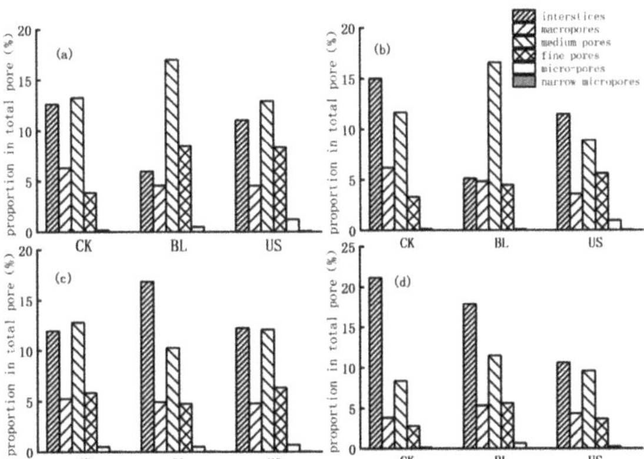

Figure 5. Distribution of soil equivalent pores in different soil layers under different treatments. (**a–d**) for 0–5 cm, 5–10 cm, 10–20 cm and 20–30 cm soil layer, respectively. CK: shifting sandy land; BL: bare land without vegetation cover in the shelterbelt; US: under the shelterbelt.

Soil water parameters of each treatment are shown in Table 5. In the 0–5 cm soil layer, θ_s, θ_f, θ_r, θ_a and θ_{ra} under BL and US were higher than those under CK. Compared with CK, θ_s of BL and US increased by 4.42% and 12.67%, θ_f increased by 68.9% and 70.41%, θ_r increased by 32.84% and 69.47%, θ_a increased by 87.84% and 70.97% and θ_{ra} increased by 87.73% and 70.43%, respectively. In the 5–10 cm soil layer, θ_s of CK was the largest, but θ_f, θ_r, θ_a and θ_{ra} of US and BL were higher than those of CK. In the 10–20 cm soil layer, higher θ_f, θ_r, θ_a and θ_{ra} were observed under US than in BL and CK. In the 20–30 cm soil layer, higher θ_s, θ_f, θ_r, θ_a and θ_{ra} were observed under BL than in US and CK. Meanwhile, the contents of available water in each treatment were almost equal to that of readily available water content.

Table 5. Soil water parameters calculated from the VG models.

Depth	Location	θ_s (%)	θ_f (%)	θ_r (%)	θ_a (%)	θ_{ra} (%)
0–5	CK	40.92	13.89	4.75	9.13	9.13
	BL	42.73	23.46	6.31	17.15	17.14
	US	46.11	23.67	8.05	15.61	15.56
5–10	CK	40.67	12.21	4.40	7.80	7.80
	BL	35.80	16.46	4.74	11.72	11.72
	US	34.32	14.39	3.82	10.56	10.51
10–20	CK	40.36	15.90	4.06	11.84	11.83
	BL	41.41	13.65	4.13	9.53	9.51
	US	41.20	17.51	5.14	12.38	12.35
20–30	CK	40.57	14.07	4.24	9.83	9.83
	BL	45.67	15.84	4.70	11.14	11.12
	US	39.59	12.38	4.54	8.59	8.52

Note: θ_s: saturated water content; θ_f: field capacity; θ_r: wilting coefficient; θ_a: available water content; θ_{ra}: readily available water content. (BC and G model were not included because the VG model had the best fitting effect).

As shown in Figure 6, the shapes of SWCC of each soil layer were similar, but the variations in soil moisture under per unit suction were clearly different. Under 0–10 cm and 1000–10,000 cm suctions, soil water was lost slowly with the increase in suction. However, under the suction of 10–1000 cm, the curve trended to be steep, and the soil water decreased rapidly with the increase in suction. Figure 6 clearly showed the differences in the course of the water retention curves of the three sampled areas. Soil water-holding capacity of 0–5 cm soil layer was highest under US, followed by BL and CK (Figure 6a). The difference in water-holding capacity among treatments in the remaining soil layers gradually became smaller. At the suction value corresponding to the field water-holding capacity (p_F = 1.8, i.e., 63 cm water column), the moisture of the 5–10 cm soil layer was as follows: BL ≥ US≥ CK; 10–20 cm soil layer was as follows: US ≥ CK ≥ BL; and 20–30 cm soil layer was as follows: BL ≥ CK ≥ US. When reaching the suction value corresponding to the temporary wilting coefficient (p_F = 3.8, i.e., 6309 cm water column), the water content of the 5–10 cm soil layer behaved as follows: BL ≥ CK ≥ US; the water content of the 10–20 cm soil layer behaved as follows: US ≥ BL ≥ CK; and the water content of the 20–30 cm soil layer behaved as follows: BL ≥ CK ≥ US.

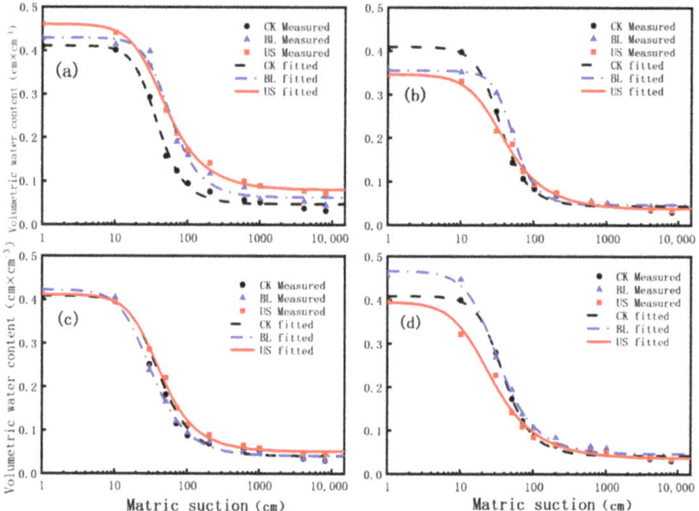

Figure 6. Water characteristic curves of each soil layer under different treatments. The fitted values are based on VG model; (**a**) 0–5 cm soil layer; (**b**) 5–10 cm soil layer; (**c**) 10–20 cm soil layer; (**d**) 20–30 cm soil layer.

4. Discussion

4.1. Artificial Shelterbelt Construction Greatly Changed the Soil Physiochemical Properties

SWCC is affected by various soil properties such as texture, bulk density, porosity, organic matter and salinity [42,53–55]. In our study, compared with shifting sandy land, the soil properties of BL and US changed obviously (Table 1), which caused the transformation of SWCC (Figure 6). Soil physiochemical properties of the 0–5 cm soil layer were changed most significantly. Shelterbelt construction under saline irrigation significantly decreased the soil bulk density, increased EC and organic matter content, and the soil texture changed from sandy soil to loamy sand and sandy loam (Table 1). After shelterbelt construction, the soil bulk density of 0–5 cm decreased significantly as compared with CK, which is primarily due to the continuous input of plant litters in the surface soil that lead to loosening the soil particles [56,57]. The bulk density of BL was the highest in the 5–30 cm soil layer, indicating that long-term saline irrigation would increase the soil bulk density [58,59]. The change in soil texture was mainly reflected by the increase in silt and clay particles. The main reason is that the shelterbelts reduced the wind speed, which promotes the precipitation of sand and the accumulation of dust fall [60,61]. The second reason is the accumulation of litter and the role of microorganisms. The volume and average radius of soil macropores increased with the increase in volumetric rock fragment content [62,63]. This directly led to the reduction in large pores and increase in small pores under BL and US (Figure 5a), which improved soil water-holding capacity and changed the SWCC. In addition to the influence of rock fragments, soil macropores are also controlled by biological factors [64].

Therefore, when the content of rock fragments is similar, the distribution of soil macropores will also be different (Figure 5b–d).

Under long-term saline irrigation, the salts were added into the soil and resulted in surface accumulation due to strong evaporation rate [13,65]. Therefore, the soil ECs of BL and US in the 0–10 cm soil layer were significantly higher than shifting sandy land, and the EC of BL and US decreased with the increasing soil depth. Vegetation cover resulted in less evaporation under US than BL, and more intense surface salt accumulation under BL. Therefore, soil EC of BL was higher than US in the 0–5 cm soil layer, and EC of BL was lower than US in the 0–10 cm soil layer. Wind erosion has an important impact on the cycle of soil organic carbon. The fine particles in the sand and dust adsorb soil organic carbon. Under the action of wind erosion, soil organic carbon is redistributed along with the movement of sand and dust [65,66]. Therefore, the increase in soil organic matter was mainly due to the accumulation of litters, and atmospheric dustfall also played a certain role in promoting it [67].

4.2. Artificial Shelterbelt Construction Increased the Soil Water-Holding Capacity

Soil water moves in pores and its transfer rate is directly determined by the size and distribution of pores. Soil bulk density is negatively correlated with soil porosity. Changes in soil primary particles such as sand, silt and clay also affect the distribution of pores. Organic matter and salinity have a direct impact on soil structure and adsorption. These soil physical properties may directly or indirectly affect soil water conductivity. Therefore, SWCC is affected by bulk density, texture, organic matter, porosity, aggregate stability, salinity and other properties [19,30–33,40–42].

Compared with CK for the 0–5 cm soil layer, vegetation coverage resulted in the decline in bulk density and increase in the salinity, organic matter and silt content under BL and US. On the one hand, salt contents in soil will occupy the pore space, and it will cause some soil particles to flocculate together, increase soil pores and enhance soil water-holding capacity [68]. Plant litters will reduce soil bulk density and increase soil organic matter content, soil saturated water content and water conductivity [69,70]. Therefore, the water contents of BL and US were higher than those of CK under the same suction, and with the increase in suction, the water contents of BL and US decreased, which means the SWCC integrally moved upwards and the trend slowed down. The contents of soil interstices and macropores in the 5–10 cm of CK were more than that in BL and US (Figure 5b). SWCC reflects the dehumidification process of soils, during which water is stored in pores and water retained in macropores is preferentially expelled as suction increases. Therefore, the change in water content per unit suction of CK was higher than that of BL and US. Therefore, the curve was the maximum under 0–10 cm suction, and then decreased rapidly to the minimum. The soil physiochemical properties of each treatment in the 10–20 cm soil layer had little difference, and their water-holding capacities were similar. For the 20–30 cm soil layer, the bulk density of BL was significantly higher than that of CK and US, and the water-holding curve was at its peak. Our results are consistent with Lipiec et al. [54]. Comparing the soil water changes over the entire suction range, the water-holding capacity of the surface 0–5 cm soil increased the most significantly, and the water-holding capacity of BL and US increased over the entire suction range compared to CK. The increase in water retention capacity of shelterbelt soils under long-term saline water irrigation is mainly due to the following reasons. Soil texture controls the physical, hydrological and chemical properties of the soil and has a strong influence on water flow paths, residence times and the magnitude and location of salt accumulation. Soil texture governs the water and solute transport. Under irrigation conditions, finer soils limit water infiltration, and coarse-grained soils retain significantly less water than fine-grained soils [71,72]. The accumulation of soil salts is dominated by sodium salts originated from the irrigation water. Excessive concentration of sodium ions in the soil solution will disperse and swell the soil structure, leading to the reduction and blockage of connected pores, thus reducing the permeability and hydraulic conductivity of the soil [73,74]. The salinity of pore water also affects the development of the diffusion double layer (DDL) around soil particles, which controls the microstructural changes in soil particles during hydration. As the salt content in irrigation water increases, the interlayer space between DDLs expands and soil aggregates are disrupted by swelling and clay dispersion [41,75]. The more significant salinity damages the soil structure and the salt stress results in higher absorption of soil water [76].

4.3. Screening of SWCC Models for Artificial Shelterbelt

The fitting R^2 results of VG model and BC model were significantly higher than those of the G model, and the *RE* was generally lower than that of G model. Therefore, the fitting results of the VG and BC models were better for SWCC, while the BC model had better fitting effect for CK. However, the fitting effects of the VG model were better than the BC model for bare BL and US, especially for the surface soil. This is because the VG model considers more influencing factors when predicting. Therefore, it has a higher accuracy on soils with more complex physiochemical properties. The soil salt content and organic matter content of BL and US were maximum, so the fitting effect of the VG model is better than the BC model. Therefore, comparing the fitting results of the three models, the results concluded that the VG model is the best choice in this regard.

When using RETC software to predict the parameters of the VG model, only the influence of bulk density and texture can be considered. In our study, it can be found that the increase in organic matter and salinity reduce the fitting accuracy of the VG model, and through grey correlation analysis, we found that the soil bulk density, sand content, pH, clay content, organic matter, silt content, EC and other physiochemical indicators of the study area had a larger correlation degree with the parameters a and n of the VG model (Table 4). Meanwhile, the basic parameters such as organic matter, pH and EC were used as input variables to predict the parameters of the VG model, and our results found that the fitting effect of the VG model could be improved [77,78]. Therefore, pH, EC, organic matter and other indicators should also be used as input variables for calculation when predicting the parameters of the VG model.

5. Conclusions

The construction of an artificial shelterbelt with long-term saline water irrigation increased the water retention capacity and the content of fine pores in the soils, and reduced the content of macropores. Artificial shelterbelt construction had the greatest impact on the surface soil, and the impact gradually decreased with the increasing soil depth. Compared with the shifting sandy land (CK) in the 0–5 cm soil layer, the saturated water content θ_s of the soil under BL and US increased by 4.42% and 12.67%, the field capacity θ_f increased by 68.9% and 70.41%, and the available water content θ_a increased by 87.84% and 70.97%. In the 5–10 cm soil layer, the θ_f, θ_r, θ_a and θ_{ra} of US and BL were higher than those of CK. In the 10–20 cm soil layer, the soil of US had the best water-holding performance. In the 20–30 cm soil layer, the soil water-holding performance of BL was the best. The available water content of each treatment was in the form of readily available water content. Comparing the coefficient of determination (R^2), root mean square error (*RSME*) and relative error (*RE*) of the three models, it was found that the G model always overestimated the soil water content and had a lower prediction accuracy, while the BC and VG models had higher prediction accuracies. Although both BC and VG models are suitable for fitting SWCC of the shelterbelt, the VG model is more effective. The parameters a and n of the VG model had a higher degree of correlation with the soil EC and organic matter. In summary, the increase in organic matter and salinity reduced the fitting accuracy of the model. When predicting the parameters of the VG model and establishing the soil transfer function, soil EC, organic matter and other indicators should be calculated as input variables.

Author Contributions: Conceptualization, C.M. and J.Z.; Data curation, L.T., M.T.J., Q.C. and J.F.; Formal analysis, C.M. and W.C.; Investigation, C.M., X.L., Q.C. and J.F.; Methodology, C.M., J.Z. and X.L.; Writing—original draft, C.M.; Writing—review and editing, J.Z. All authors have read and agreed to the published version of the manuscript.

Funding: This study was supported by the National Natural Science Foundation of China (No. 41877541, 41471222), Key Scientific and Technological Project of Shaanxi Province (2022ZDLNY02-03).

Institutional Review Board Statement: Not applicable.

Informed Consent Statement: Not applicable.

Data Availability Statement: All data reported here are available from the authors upon request.

Conflicts of Interest: The authors declare no conflict of interest.

References

1. Kaisermann, A.; de Vries, F.T.; Griffiths, R.I.; Bardgett, R.D. Legacy effects of drought on plant–soil feedbacks and plant–plant interactions. *New Phytol.* **2017**, *215*, 1413–1424. [CrossRef]
2. Zhang, Q.; Shao, M.; Jia, X.; Wei, X. Changes in soil physical and chemical properties after short drought stress in semi-humid forests. *Geoderma* **2019**, *338*, 170–177. [CrossRef]
3. Mantovani, D.; Veste, M.; Boldt-Burisch, K.; Fritsch, S.; Koning, L.A.; Freese, D. Carbon allocation, nodulation, and biological nitrogen fixation of black locust (*Robinia pseudoacacia* L.) under soil water limitation. *Ann. For. Res.* **2015**, *58*, 259–274. [CrossRef]
4. Yu, T.; Feng, Q.; Si, J.; Xi, H.; Li, Z.; Chen, A. Hydraulic redistribution of soil water by roots of two desert riparian phreatophytes in northwest China's extremely arid region. *Plant Soil* **2013**, *372*, 297–308. [CrossRef]
5. Deng, L.; Wang, K.; Li, J.; Zhao, G.; Shang, G.Z. Effect of soil moisture and atmospheric humidity on both plant productivity and diversity of native grasslands across the Loess Plateau, China. *Ecol. Eng.* **2016**, *94*, 525–531. [CrossRef]
6. Li, X.R.; Ma, F.Y.; Xiao, H.L.; Wang, X.P.; Kim, K.C. Long-term effects of revegetation on soil water content of sand dunes in arid region of Northern China. *J. Arid Environ.* **2004**, *57*, 1–16. [CrossRef]
7. Lei, J.Q.; Li, S.Y.; Jin, Z.Z.; Fan, J.L.; Wang, H.F.; Fan, D.D.; Zhou, H.W.; Gu, F.; Qiu, Y.Z.; Xu, B. Comprehensive eco-environmental effects of the shelter-forest ecological engineering along the Tarim Desert Highway. *Chin. Sci. Bull.* **2008**, *53*, 190–202. [CrossRef]
8. Li, C.; Shi, X.; Mohamad, O.A.; Gao, J.; Xu, X.; Xie, Y. Moderate irrigation intervals facilitate establishment of two desert shrubs in the Taklimakan Desert Highway Shelterbelt in China. *PLoS ONE* **2017**, *12*, e0180875. [CrossRef]
9. Dai, Y.; Zheng, X.J.; Tang, L.S.; Li, Y. Stable oxygen isotopes reveal distinct water use patterns of two Haloxylon species in the Gurbantonggut Desert. *Plant Soil* **2015**, *389*, 73–87. [CrossRef]
10. Zhang, J.G.; Xu, X.W.; Lei, J.Q.; Sun, S.G.; Fan, J.L.; Li, S.Y.; Qiu, Y.Z.; Xu, B. The salt accumulation at the shifting aeolian sandy soil surface with high salinity groundwater drip irrigation in the hinterland of the Taklimakan Desert. *Chin. Sci. Bull.* **2008**, *53*, 63–70. [CrossRef]
11. Feikema, P.M.; Morris, J.D.; Connell, L.D. The water balance and water sources of a Eucalyptus plantation over shallow saline groundwater. *Plant Soil* **2010**, *332*, 429–449. [CrossRef]
12. Zhang, J.; Xu, X.; Li, S.; Zhao, Y.; Zhang, A.; Zhang, T.; Jiang, R. Is the Taklimakan Desert Highway Shelterbelt Sustainable to Long-Term Drip Irrigation with High Saline Groundwater? *PLoS ONE* **2016**, *11*, e0164106. [CrossRef]
13. Li, C.; Lei, J.; Zhao, Y.; Xu, X.; Li, S. Effect of saline water irrigation on soil development and plant growth in the Taklimakan Desert Highway shelterbelt. *Soil Tillage Res.* **2015**, *146*, 99–107. [CrossRef]
14. Wang, R.; Kang, Y.; Wan, S.; Hu, W.; Liu, S.; Liu, S. Salt distribution and the growth of cotton under different drip irrigation regimes in a saline area. *Agric. Water Manag.* **2011**, *100*, 58–69. [CrossRef]
15. Zhang, J.G.; Xu, X.W.; Lei, J.Q.; Li, S.Y.; Hill, R.L.; Zhao, Y. The effects of soil salt crusts on soil evaporation and chemical changes in different ages of Taklimakan Desert Shelterbelts. *J. Soil Sci. Plant Nutr.* **2013**, *13*, 1019–1028. [CrossRef]
16. Likos, W.J.; Yao, J. Effects of constraints on van Genuchten parameters for modeling soil-water characteristic curves. *J. Geotech. Geoenviron. Eng.* **2014**, *140*, 06014013. [CrossRef]
17. Guo, Z.Q.; Lai, Y.M.; Jin, J.F.; Zhou, J.R.; Sun, Z.; Zhao, K. Effect of Particle Size and Solution Leaching on Water Retention Behavior of Ion-Absorbed Rare Earth. *Geofluids* **2020**, *2020*, 4921807. [CrossRef]
18. Raghuram, A.S.S.; Basha, B.M.; Moghal, A.A.B. Effect of Fines Content on the Hysteretic Behavior of Water-Retention Characteristic Curves of Reconstituted Soils. *J. Mater. Civ. Eng.* **2020**, *32*, 04020057. [CrossRef]
19. Wang, M.; Liu, H.; Lennartz, B. Small-scale spatial variability of hydro-physical properties of natural and degraded peat soils. *Geoderma* **2021**, *399*, 115123. [CrossRef]
20. Kalhoro, S.A.; Xu, X.; Ding, K.; Chen, W.; Shar, A.G.; Rashid, M. The effects of different land uses on soil hydraulic properties in the Loess Plateau, Northern China. *Land Degrad. Dev.* **2018**, *29*, 3907–3916. [CrossRef]
21. Zhou, H.; Mooney, S.J.; Peng, X. Bimodal Soil Pore Structure Investigated by a Combined Soil Water Retention Curve and X-Ray Computed Tomography Approach. *Soil Sci. Soc. Am. J.* **2017**, *81*, 1270–1278. [CrossRef]

22. Alves, R.D.; Gitirana, G.F.N., Jr.; Vanapalli, S.K. Advances in the modeling of the soil–water characteristic curve using pore-scale analysis. *Comput. Geotech.* **2020**, *127*, 103766. [CrossRef]
23. Guo, Z.Q.; Zhou, J.R.; Zhou, K.F.; Jin, J.F.; Wang, X.J.; Kui, Z. Soil-water characteristics of weathered crust elution-deposited rare earth ores. *Trans. Nonferrous Met. Soc. China* **2021**, *31*, 1452–1464. [CrossRef]
24. Zhao, G.; Zou, W.; Han, Z.; Wang, D.X.; Wang, X.Q. Evolution of soil-water and shrinkage characteristics of an expansive clay during freeze-thaw and drying-wetting cycles. *Cold Reg. Sci. Technol.* **2021**, *186*, 103275. [CrossRef]
25. Wu, T.N.; Wu, G.L.; Wang, D.; Shi, Z.H. Soil-hydrological properties response to grazing exclusion in a steppe grassland of the Loess Plateau. *Environ. Earth Sci.* **2013**, *71*, 745–752. [CrossRef]
26. Matlan, S.J.; Taha, M.R.; Mukhlisin, M. Assessment of model consistency for determination of soil–water characteristic curves. *Arab. J. Sci. Eng.* **2016**, *41*, 1233–1240. [CrossRef]
27. Genuchten, V.; Th, M.A. Closed-form equation for predicting the hydraulic conductivity of unsaturated soils. *Soil Sci. Soc. Am. J.* **1980**, *44*, 892–898. [CrossRef]
28. Gardner, W.R. Some steady-state solutions of the unsaturated moisture flow equation with application to evaporation from a water table. *Soil Sci.* **1958**, *85*, 228–232. [CrossRef]
29. Brooks, R.H.; Corey, A.T. *Hydraulic Properties of Porous Media*; Colorado State University Hydrology Paper No. 3 (March); Colorado State University: Fort Collins, CO, USA, 1964.
30. Libutti, A.; Francavilla, M.; Monteleone, M. Hydrological Properties of a Clay Loam Soil as Affected by Biochar Application in a Pot Experiment. *Agronomy* **2021**, *11*, 489. [CrossRef]
31. John, A.; Fuentes, H.R.; George, F. Characterization of the water retention curves of Everglades wetland soils. *Geoderma* **2021**, *381*, 114724. [CrossRef]
32. Gao, Z.; Hu, X.; Li, X.Y. Changes in soil water retention and content during shrub encroachment process in Inner Mongolia, northern China. *Catena* **2021**, *206*, 105528. [CrossRef]
33. Cao, B.; Tian, Y.; Gui, R.; Liu, Y. Experimental Study on the Effect of Key Factors on the Soil–Water Characteristic Curves of Fine-Grained Tailings. *Front. Environ. Sci.* **2021**, *9*, 710986. [CrossRef]
34. Ren, X.; Kang, J.; Ren, J.; Chen, X.; Zhang, M. A method for estimating soil water characteristic curve with limited experimental data. *Geoderma* **2020**, *360*, 114013. [CrossRef]
35. Andrabi, S.G.; Ghazanfari, E.; Vahedifard, F. An empirical relationship between Brooks–Corey and Fredlund–Xing soil water retention models. *J. Porous Media* **2019**, *22*, 1423–1437. [CrossRef]
36. Jahara, M.S.; Muhammad, M.; Raihan, T.M. Performance evaluation of four-parameter models of the soil-water characteristic curve. *Sci. World J.* **2014**, *2014*, 569851.
37. Zhou, J.; Yu, J.L. Influences affecting the soil-water characteristic curve. *J. Zhejiang Univ. Sci. A* **2005**, *6*, 797–804. [CrossRef]
38. Lai, J.; Wang, Q. Comparison of Soil Water Retention Curve Model. *J. Soil Water Conserv.* **2003**, *17*, 137–140.
39. Dang, M.; Chai, J.; Xu, Z.; Qin, Y.; Cao, J.; Liu, F. Soil water characteristic curve test and saturated-unsaturated seepage analysis in Jiangcungou municipal solid waste landfill, China. *Eng. Geol.* **2020**, *264*, 105374. [CrossRef]
40. Jiang, X.; Wu, L.; Wei, Y. Influence of Fine Content on the Soil–Water Characteristic Curve of Unsaturated Soils. *Geotech. Geol. Eng.* **2019**, *38*, 1371–1378. [CrossRef]
41. Lu, Y.; Abuel-Naga, H.; Leong, E.C.; Bouazza, A.; Lock, P. Effect of water salinity on the water retention curve of geosynthetic clay liners. *Geotext. Geomembr.* **2018**, *46*, 707–714. [CrossRef]
42. Xing, X.; Ma, X. Differences in loam water retention and shrinkage behavior: Effects of various types and concentrations of salt ions. *Soil Tillage Res.* **2017**, *167*, 61–72. [CrossRef]
43. Dong, Y.Y.; Zhao, C.; Yu, Z.T.; Wang, D.D.; Ban, C.G. Characteristic Curves and Models Analysis of Soil Water in Interdune at the Southern Edge of Gurbantunggut Desert. *J. Soil Water Conserv.* **2017**, *31*, 166–171. (In Chinese with English Abstract)
44. Ding, X.Y.; Zhou, Z.B.; Lei, J.Q.; Wang, Y.D.; Lu, J.J.; Li, X.J. Analysis and comparison of models for soil water characteristic curves of Tarim Desert Highway Shelterbelt. *Arid Land Geogr.* **2015**, *38*, 985–993. (In Chinese with English Abstract)
45. Bayat, H.; Ebrahimi, E. Effects of various input levels and different soil water retention curve models on water content estimation using different statistical methods. *Hydrol. Res.* **2016**, *47*, 312–332. [CrossRef]
46. Li, S.; Xie, Y.; Xin, Y.; Liu, G.; Wang, W.; Gao, X.; Zhai, J.R.; Li, J. Validation and Modification of the Van Genuchten Model for Eroded Black Soil in Northeastern China. *Water* **2020**, *12*, 2678. [CrossRef]
47. Sun, F.; Xiao, B.; Li, S.; Kidron, G.J. Towards moss biocrust effects on surface soil water holding capacity: Soil water retention curve analysis and modeling. *Geoderma* **2021**, *399*, 115120. [CrossRef]
48. Soil Survey Staff. *Kellogg Soil Survey Laboratory Methods Manual*; Soil Survey Investigations Report No. 42, Version 5.0; Burt, R., Soil Survey Staff, Eds.; USDA Natural Resources Conservation Service: Lincoln, NE, USA, 2014.
49. Chen, J.; Chai, H.; Gillerman, L.; Liu, C.; Zhang, L. Impact of treated waste water quality on repellent and wettable soil water characteristic curve. *Trans. Chin. Soc. Agric. Eng.* **2018**, *34*, 121–127. (In Chinese with English Abstract)
50. Gao, H.; Guo, S.; Liu, W.; Li, M.; Zhang, J. Spatial Variability of Soil Water Retention Curve under Fertilization Practices in Arid-highland of the Loess Plateau. *Trans. Chin. Soc. Agric. Mach.* **2014**, *45*, 161–165+176. (In Chinese with English Abstract)
51. Zhang, J.; Wang, Q.; Wang, W.; Zhang, X. The dispersion mechanism of dispersive seasonally frozen soil in western Jilin Province. *Bull. Eng. Geol. Environ.* **2021**, *80*, 5493–5503. [CrossRef]

52. Li, S.; Lu, L.; Gao, Y.; Zhang, Y.; Shen, D. An Analysis on the Characteristics and Influence Factors of Soil Salinity in the Wasteland of the Kashgar River Basin. *Sustainability* **2022**, *14*, 3500. [CrossRef]
53. Zhang, G.S.; Chan, K.Y.; Oates, A.; Heenan, D.P.; Huang, G.B. Relationship between soil structure and runoff/soil loss after 24 years of conservation tillage. *Soil Tillage Res.* **2007**, *92*, 122–128. [CrossRef]
54. Lipiec, J.; Walczak, R.; Witkowska-Walczak, B.; Nosalewicz, A.; Słowińska-Jurkiewicz, A.; Sławiński, C. The effect of aggregate size on water retention and pore structure of two silt loam soils of different genesis. *Soil Tillage Res.* **2007**, *97*, 239–246. [CrossRef]
55. Lan, Z.; Zhao, Y.; Zhang, J.; Yang, X.; Sial, T.A.; Khan, M.N. Effects of the long-term fertilization on pore and physicochemical characteristics of loess soil in Northwest China. *Agron. J.* **2020**, *112*, 4741–4751. [CrossRef]
56. Zhang, J.; Zuo, X.; Zhou, X.; Lv, P.; Lian, J.; Yue, X. Long-term grazing effects on vegetation characteristics and soil properties in a semiarid grassland, northern China. *Environ. Monit. Assess.* **2017**, *189*, 216. [CrossRef]
57. Keesstra, S.; Pereira, P.; Novara, A.; Brevik, E.C.; Azorin-Molina, C.; Parras-Alcántara, L.; Jordán, A.; Cerdà, A. Effects of soil management techniques on soil water erosion in apricot orchards. *Sci. Total Environ.* **2016**, *551*, 357–366. [CrossRef] [PubMed]
58. Yuan, C.; Feng, S.; Wang, J.; Huo, Z.; Ji, Q. Effects of irrigation water salinity on soil salt content distribution, soil physical properties and water use efficiency of maize for seed production in arid Northwest China. *Int. J. Agric. Biol. Eng.* **2018**, *11*, 137–145. [CrossRef]
59. Cheng, M.; Wang, H.; Fan, J.; Wang, X.; Sun, X.; Yang, L.; Zhang, S.; Xiang, Y.; Zhang, F. Crop yield and water productivity under salty water irrigation: A global meta-analysis. *Agric. Water Manag.* **2021**, *256*, 107105. [CrossRef]
60. Jia, X.H.; Xin-Rong, L.I.; Wang, X.P.; Fan, H.W.; Zhao, J.L. Primary study of spatial heteroge-neity of soil property in processes of shifting sand fixation in south-easter Tengger Desert. *J. Soil Water Conserv.* **2003**, *17*, 46–49. (In Chinese with English Abstract)
61. Zhang, S.; Ding, G.D.; Yu, M.H.; Gao, G.L.; Zhao, Y.Y.; Wu, G.H.; Wang, L. Effect of straw checkerboards on wind proofing, sand fixation, and ecological restoration in shifting sandy land. *Int. J. Environ. Res. Public Health* **2018**, *15*, 2184. [CrossRef]
62. Shi, Z.J.; Xu, L.H.; Wang, Y.H.; Yang, X.H.; Jia, Z.Q.; Guo, H.; Xiong, W.; Yu, P.T. Effect of rock fragments on macropores and water effluent in a forest soil in the stony mountains of the Loess Plateau, China. *Afr. J. Biotechnol.* **2012**, *11*, 9350–9361.
63. Xu, L.H.; Shi, Z.J.; Wang, Y.H.; Chu, X.Z.; Yu, P.T.; Xiong, W. Contribution of rock fragments on formation of forest soil macropores in the stoney mountains of the Loess Plateau, China. *J. Food Agric. Environ.* **2012**, *10*, 1220–1226.
64. Wang, Z.; Li, G.; Li, X.; Shan, S.; Zhang, J.; Li, S.; Fan, J. Characteristics of moisture and salinity of soil in Taklimakan Desert, China. *Water Sci. Technol.* **2012**, *66*, 1162–1170. [CrossRef] [PubMed]
65. Webb, N.P.; Strong, C.L.; Chappell, A.; Marx, S.K.; McTainsh, G.H. Soil organic carbon enrichment of dust emissions: Magnitude, mechanisms and its implications for the carbon cycle. *Earth Surf. Process. Landf.* **2013**, *38*, 1662–1671. [CrossRef]
66. Du, H.; Li, S.; Webb, N.P.; Zuo, X.; Liu, X. Soil organic carbon (SOC) enrichment in aeolian sediments and SOC loss by dust emission in the desert steppe, China. *Sci. Total Environ.* **2021**, *798*, 149189. [CrossRef] [PubMed]
67. Chen, S.Y.; Huang, J.P.; Li, J.X.; Jia, R.; Jiang, N.X.; Kang, L.T.; Ma, X.J.; Xie, T.T. Comparison of dust emissions, transport, and deposition between the Taklimakan Desert and Gobi Desert from 2007 to 2011. *Sci. China Earth Sci.* **2017**, *60*, 1338–1355. [CrossRef]
68. Hu, C.; Wang, H.; Wu, Y.; Lu, J.; Liu, C. Difference of Soil Water Characteristic Curves of Subtropical Soils with NaCl Solutions Treatments and Models Optimization. *Trans. Chin. Soc. Agric. Mach.* **2018**, *49*, 290–296+329. (In Chinese with English Abstract)
69. Wei, Y.; Wang, H.; Yu, W.U.; Jia, R.; Jiang, N.; Kang, L.; Ma, X.; Xie, T. Effect of Biochar on Soil Hydrodynamic Parameters under Different Slopes. *Trans. Chin. Soc. Agric. Mach.* **2019**, *50*, 231–240. (In Chinese with English Abstract)
70. Sun, X.; Fang, K.; Fei, Y.; She, D. Structure and Hydraulic Characteristics of Saline Soil Improved by Applying Biochar Based on Micro-CT Scanning. *Trans. Chin. Soc. Agric. Mach.* **2019**, *50*, 242–249. (In Chinese with English Abstract)
71. Ortiz, A.C.; Jin, L. Chemical and hydrological controls on salt accumulation in irrigated soils of southwestern U.S. *Geoderma* **2021**, *391*, 114976. [CrossRef]
72. Cox, C.; Jin, L.; Ganjegunte, G.; Borrok, D.; Lougheed, V.; Ma, L. Soil quality changes due to flood irrigation in agricultural fields along the Rio Grande in western Texas. *Appl. Geochem.* **2018**, *90*, 87–100. [CrossRef]
73. Ferretti, G.; Di, G.D.; Faccini, B.; Coltorti, M. Mitigation of sodium risk in a sandy agricultural soil by the use of natural zeolites. *Environ. Monit. Assess.* **2018**, *190*, 646. [CrossRef]
74. Qi, Z.; Feng, H.; Zhao, Y.; Zhang, T.; Yang, A.; Zhang, Z. Spatial distribution and simulation of soil moisture and salinity under mulched drip irrigation combined with tillage in an arid saline irrigation district, northwest China. *Agric. Water Manag.* **2018**, *201*, 219–231. [CrossRef]
75. Thyagaraj, T.; Salini, U. Effect of pore fluid osmotic suction on matric and total suctions of compacted clay. *Geotechnique* **2015**, *65*, 952–960. [CrossRef]
76. Yang, G.; Liu, S.; Yan, K.; Tian, L.; Li, P.; Li, X.; He, X. Effect of Drip Irrigation with Brackish Water on the Soil Chemical Properties for a Typical Desert Plant (Haloxylon Ammodendron) in the Manas River Basin. *Irrig. Drain.* **2020**, *69*, 460–471. [CrossRef]
77. Abbasi, Y.; Ghanbarian-Alavijeh, B.; Liaghat, A.M. Evaluation of Pedotransfer Functions for Estimating Soil Water Retention Curve of Saline and Saline-Alkali Soils of Iran. *Pedosphere* **2011**, *21*, 230–237. [CrossRef]
78. Nuth, M.; Laloui, L. Advances in modelling hysteretic water retention curve in deformable soils. *Comput. Geotech.* **2008**, *35*, 835–844. [CrossRef]

Article

Spatial Heterogeneity and Driving Factors of Soil Moisture in Alpine Desert Using the Geographical Detector Method

Zhiwei Zhang [1,2], Huiyan Yin [1,2], Ying Zhao [3], Shaoping Wang [4], Jiahua Han [1,2], Bo Yu [1,2] and Jie Xue [5,6,7,*]

1 College of Resources and Environmental Sciences, Tibet Agriculture & Animal Husbandry University, Nyingchi 860000, China; aiwoweige@163.com (Z.Z.); huiyanyin@163.com (H.Y.); hjh17865664171@163.com (J.H.); yb18005648616@163.com (B.Y.)
2 Key Laboratory of Forest Ecology in Tibet Plateau, Tibet Agricultural & Animal Husbandry University, Nyingchi 860000, China
3 College of Resources and Environmental Engineering, Ludong University, Yantai 264025, China; yzhaosoils@gmail.com
4 State Key Laboratory of Cryospheric Science, Northwest Institute of Eco-Environment and Resources, Chinese Academy of Sciences, Lanzhou 730000, China; w1215lin@163.com
5 State Key Laboratory of Desert and Oasis Ecology, Xinjiang Institute of Ecology and Geography, Chinese Academy of Sciences, Urumqi 830011, China
6 Cele National Station of Observation and Research for Desert-Grassland Ecosystems, Cele, Hetian 848300, China
7 University of Chinese Academy of Sciences, Beijing 100049, China
* Correspondence: xuejie11@mails.ucas.ac.cn

Citation: Zhang, Z.; Yin, H.; Zhao, Y.; Wang, S.; Han, J.; Yu, B.; Xue, J. Spatial Heterogeneity and Driving Factors of Soil Moisture in Alpine Desert Using the Geographical Detector Method. *Water* **2021**, *13*, 2652. https://doi.org/10.3390/w13192652

Academic Editor: Alexander Yakirevich

Received: 8 August 2021
Accepted: 22 September 2021
Published: 26 September 2021

Abstract: Soil moisture is a vital factor affecting the hydrological cycle and the evolution of soil and geomorphology, determining the formation and development of the vegetation ecosystem. The previous studies mainly focused on the effects of different land use patterns and vegetation types on soil hydrological changes worldwide. However, the spatial heterogeneity and driving factors of soil gravimetric water content in alpine regions are seldom studied. On the basis of soil sample collection, combined with geostatistical analysis and the geographical detector method, this study examines the spatial heterogeneity and driving factors of soil gravimetric water content in the typical alpine valley desert of the Qinghai–Tibet Plateau. Results show that the average value of soil gravimetric water content at different depths ranges from 3.68% to 7.84%. The optimal theoretical models of soil gravimetric water content in 0–50 cm layers of the dune are different. The nugget coefficient shows that the soil gravimetric water content in the dune has a strong spatial correlation at different depths, and the range of the optimal theoretical model of semi-variance function is 31.23–63.38 m, which is much larger than the 15 m spacing used for sampling. The ranking of the influence of each evaluation factor on the alpine dune is elevation > slope > location > vegetation > aspect. The interaction detection of factors indicates that an interaction exists among evaluation factors, and no factors are independent of one another. In each soil layer of 0–50 cm, the interaction among evaluation factors has a two-factor enhancement and a nonlinear enhancement effect on soil gravimetric water content. This study contributes to the understanding of spatial heterogeneity and driving factors of soil moisture in alpine deserts, and guidance of artificial vegetation restoration and soil structure analysis of different desert types in alpine cold desert regions.

Keywords: geographical detector; alpine dunes; spatial heterogeneity; soil moisture; Qinghai–Tibet Plateau

1. Introduction

Soil moisture is a vital factor affecting the hydrological cycle and the evolution of soil and geomorphology, determining the formation and development of the vegetation ecosystem, especially in alpine deserts [1]. The soil moisture distribution directly controls the carrying capacity of vegetation and the restoration and reconstruction of degraded ecosystems. However, the soil moisture distribution has high variability both spatially and temporally [2]. The spatial heterogeneity limits the understanding of the spatial pattern of soil moisture in water resource management and ecosystem restoration [3–6]. Therefore, examining the spatial heterogeneity of soil moisture and its response to environmental factors is important in ensuring the sustainability of alpine ecosystems.

In recent decades, research on soil moisture has permeated various ecological and environmental fields throughout the world, especially in the arid and semiarid areas. It was reported that the change of soil moisture in sandy soils is affected not only by topography and vegetation but also by the spatiotemporal variation of precipitation with a strong dependency. Meanwhile, the previous studies have mainly focused on the effects of different land use patterns, vegetation types, spatiotemporal factors, and soil properties on soil hydrological changes in sandy soils [7–13].

In addition, the influencing factors of the spatial heterogeneity of soil water in arid sandy soils have also been examined [14–16]. For example, ref [17] measured the soil gravimetric water content and analyzed the observation data in the Yarlung Zangbu River. Ref [18] identified the spatial patterns and temporal stability of topsoil gravimetric water content in a Mediterranean fallow cereal field. Ref. [19] analyzed the spatial distribution characteristics of soil gravimetric water content in the southern part of Iran. With respect to the spatial heterogeneity of soil water, a distributed parameter model has been proposed to link soil gravimetric water content as the system input to describe the spatiotemporal variability of soil gravimetric water content in arid regions [20]. Nevertheless, the spatial heterogeneity and driving factors of soil gravimetric water content in alpine cold deserts is seldom studied.

The geographical detector is an effective approach to detecting spatial differences and revealing the driving factors [21]. Many scholars have studied the spatial heterogeneity of soil gravimetric water content and its influencing factors through correlational analysis, stepwise linear regression, and principal component analysis [1]. However, few of them have used the superiority of the geographical detector method to explore spatial heterogeneity and driving factors.

The objective of this study is to elucidate the spatial heterogeneity of soil gravimetric water content and its driving factors in the alpine valley desert of the Qinghai–Tibet Plateau. Geostatistical analysis is conducted to detect the spatial heterogeneity of soil moisture. The Kriging method is used to draw the contour distribution map of soil moisture at different depths. The influences of different factors on the soil moisture changes in alpine dunes are quantified using the geographical detector method. The results of this study clarify the main influencing factors of soil moisture, the understanding of spatial heterogeneity, and the guidance of artificial vegetation restoration and soil structure analysis of different desert types in alpine cold desert regions.

2. Materials and Methods

2.1. Study Area and Data Sources

The study area is located in the northern part of the Himalaya mountains, the upper reaches of the Yiruzangpo River Basin, and Kamba County of Tibet, China. It covers a latitude of 28°21′25.51″ N to 28°24′8.66″ N and a longitude of 88°25′9.63″ E to 88°28′12.55″ E (Figure 1). The elevation is between 4418 and 4692 m. The area is a representative alpine desert steppe and sandy desertification land, belonging to the semiarid climate of the plateau sub-cold monsoon [22]. It is characterized by extreme coldness, with an annual mean temperature of 0.4 °C and an accumulated precipitation of 434.3 mm (Figure 1). The soil is classified as fluvisols, gleysols, and arenosols by the universal soil classification system [23]. The soil bulk density is approximately 1.53 g/cm^3. Minimal shrubs are distributed in the wide valley of the study area.

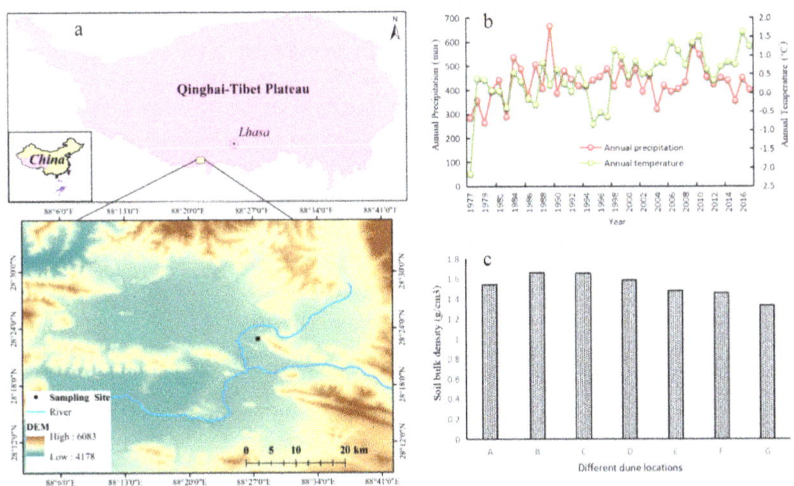

Figure 1. Overview of the study area. (**a**) location and distribution of soil sampling points, (**b**) annual mean temperature and annual precipitation, (**c**) soil bulk density in the study area.

A representative sand dune which runs from north to south was chosen as the study location to examine the spatial change of soil moisture of the alpine desert. The length of the dune is approximately 500 m from north to south and 117 m from east to west. The height is approximately 13.5 m. The west of the dune is a wide valley, and the east adjoins the mountain. The soil samples were collected on 23 September 2020, covering the bottom of the dune to the leeward slope. The sample plot size is 100 × 100 m. A grid spacing of approximately 15 × 15 m was used to lay 60 sampling points. The locations of the sampling points were recorded by GPS (Figure 2). The sampling depths are 0–10, 10–20, 20–30, 30–40, and 40–50 cm. The soil samples of each layer were evenly taken from top to bottom. Three time-repetition samples were placed into three aluminum boxes, which were immediately sealed and brought back to the laboratory for soil gravimetric water content. In addition, the soil gravimetric water content of 0–50 cm layer, which is the average in 0–10, 10–20, 20–30, 30–40, and 40–50 cm layers, is used to compare the changes of the overall soil gravimetric water content with other different soil layers.

Since the spatial heterogeneity of soil gravimetric water content in alpine deserts is reflected through the soil gravimetric water content change of a 0–50 cm soil layer in a 100 × 100 m quadrat in the sand dune, the soil gravimetric water content mainly changes in different locations of the sand dune. The location, elevation, slope, aspect, and vegetation coverage are the main factors controlling the soil water pattern [24]. Therefore, this study selected the location of dunes, elevation, slope, aspect, and vegetation coverage as explanatory variables to examine the driving factors of soil gravimetric moisture content. In the predictor variables, the location of dunes, elevation, and aspect are measured using real-time kinematic (RTK), while the slope is determined by the gradienter measurement tool. The vegetation coverage is measured by normalized difference index (NDI), which is defined as the proportion of the vertical projection of vegetation to the total area. The vegetation coverage value of each sampling point is the average of the three repeated quadrats from digital photographs [25,26].

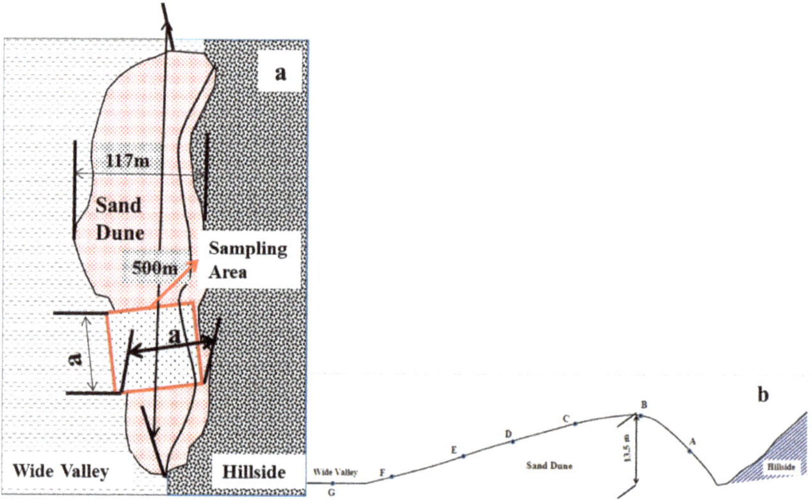

Figure 2. (**a**) Sketch of the planform of measuring positions on a sand dune (a = 100 m) and (**b**) sampling location of the dune: A is the middle part of the leeward slope, B is the top, C is the middle and upper part of the windward slope, D is the middle part of the windward slope, E is the middle and lower part of the windward slope, F is the bottom part of the windward slope, and G is the interdune land.

2.2. Methods

2.2.1. Semi-Variogram Model and Kriging Method

The semi-variogram is a function of the semi-variation value of a data point and the distance among data points. In this study, the semi-variogram was used to illustrate the graphical spatial correlation representation of soil gravimetric water content. The expression is written as [27–29]:

$$\gamma(h) = \frac{1}{2N(h)} \sum_{i=1}^{N(h)} [Z(x_i) - Z(x_i + h)]^2 \tag{1}$$

where $\gamma(h)$ is the semi-distributed function, $Z(x_i)$ and $Z(x_i + h)$ are the measured values of sampling points (x_i) and $(x_i + h)$ respectively, h is the interval distance between the sampling points, and $N(h)$ is the number of interval distances of all observation points in the study area.

To describe the variation characteristics of soil gravimetric water content, three theoretical models of the variogram were established to select the theoretical and optimal curves:

(a) Spherical model [29,30]:

$$\gamma(h) = \begin{cases} C_0 + C & (h > a) \\ C_0 + C \cdot \left[\frac{3}{2}\frac{h}{a} - \frac{1}{2}\left(\frac{h}{a}\right)^3\right] & (0 < h \le a) \\ 0 & (h = 0) \end{cases} \tag{2}$$

(b) Exponential model [29,30]:

$$\gamma(h) = \begin{cases} C_0 + C \cdot (1 - e^{-h/a}) & (h > 0) \\ 0 & (h = 0) \end{cases} \tag{3}$$

(c) Gaussian model [29,30]:

$$\gamma(h) = \begin{cases} C_0 + C \cdot (1 - e^{-h^2/a^2}) & (h > 0) \\ 0 & (h = 0) \end{cases} \quad (4)$$

where C_0 is the nugget value, C is the partial sill value, $C_0 + C$ is the sill value, a is the range value, and $C_0/(C_0 + C)$ is the nugget coefficient, which describes the strong, medium, and weak heterogeneity, at <25%, 25–75%, and >75%, respectively.

To draw the contour distribution map of soil moisture at different depths, Kriging interpolation was used to obtain spatial interpolation of the soil moisture in the study area. The general formula can be given as [31]:

$$\hat{Z}(s_0) = \sum_{i=1}^{N} \lambda_i Z(s_i) \quad (5)$$

where $Z(s)$ is the measured value at the ith location, s is the prediction location, λ is an unknown weight for the measured value at the ith location, and N is the number of measured values. The interpolation results are illustrated by the contour map.

2.2.2. Geographical Detector Method

The geographical detector is a statistical approach to assessing the impact of different environmental factors on a target variable. It is composed of four geographical detectors: factor, ecological, risk, and interaction detectors [21]. In this study, the spatial heterogeneity of soil moisture is impacted by many environmental factors. The geographical detector method is used to explore the potential impact factors versus the spatial distribution of soil moisture and to identify each explanatory variable's relative importance [32]. In this study, the elevation, slope, location, vegetation, and aspect were selected as explanatory variables for considering being the main driving factors [2,33–35].

If an explanatory factor X drives the spatial distribution of soil moisture at 0–10, 10–20, 20–30, 30–40, 40–50, and 0–50 cm depth, the spatial distribution of soil moisture is similar to that of X, which can present the pattern of soil moisture completely. The degree of spatial association between X and stratified heterogeneity of soil moisture can be quantified using the factor detector. The mathematical expression is shown by the q -statistic:

$$q = 1 - \frac{\sum_{h=1}^{L} N_h \delta_h^2}{N\delta^2} \quad (6)$$

where δ^2 is the population variance of soil moisture, δ_h^2 is the variance of stratum h, N is the number of total samples with L strata, and N_h is the number of sample units in strata h. The value of the q-statistic is within the range of 0–1. When the q value is 1, the factor X can completely explain the distribution of soil moisture, and vice versa [21].

The interactive detector can detect the interactions of multiple factors with soil moisture, namely, assess whether the mutual action of factors X_1 and X_2 could enhance or weaken the explanatory power to soil moisture, which depends on the relationship between $q(X_1 \cap X_2)$ and $q(X_1)$ or $q(X_2)$ (Table 1). The ecological detector is applied to compare whether the influences of two factors X_1 and X_2 have significant differences on the spatial distribution of soil moisture with F-statistics.

Table 1. Types of interaction between two factors [21,32,36].

Interaction	Description
Weaken, nonlinear	$q(X_1 \cap X_2) < \text{Min}\,(q(X_1), q(X_2))$
Weaken, univariate Min	$(q(X_1), q(X_2)) < q(X_1 \cap X_2) < \text{Max}\,(q(X_1), q(X_2))$
Enhance, bivariate	$q(X_1 \cap X_2) > \text{Max}\,(q(X_1), q(X_2))$
Independent	$q(X_1 \cap X_2) = q(X_1) + q(X_2)$
Enhance, nonlinear	$q(X_1 \cap X_2) > q(X_1) + q(X_2)$

3. Results

3.1. Statistical Characteristics of Soil Moisture in the Alpine Valley Desert

The statistical characteristics of soil gravimetric water content in alpine valley dunes are shown in Table 2. In accordance with the extreme and average values of soil moisture content, the soil gravimetric water content of the sampling points first increases and then decreases with the increase in soil depth. On the contrary, the coefficient of variation of soil gravimetric water content in the sampling points first decreases and then increases with the increase in soil depth. The variation coefficients of soil gravimetric water content in different soil layers are between 130% and 169%, showing strong variability.

Table 2. Descriptive statistics of soil moisture at different layers.

Depth /cm	Minimum /%	Maximum /%	Mean /%	Standard Deviation	Variation /%	Kurtosis	Skewness	K-S Test			
								Z Value *	p Value *	Z Value #	p Value #
0–10	0.07	20.93	3.68	6.23	169	2.53	2.02	0.34	0	0.82	0.54
10–20	0.27	38.33	7.06	10.71	152	2.93	2.11	0.35	0	0.83	0.40
20–30	0.51	33.03	7.84	10.28	131	0.77	1.55	0.35	0	0.80	0.24
30–40	0.59	28.85	6.63	8.60	130	0.53	1.49	0.36	0	0.80	0.31
40–50	0.26	21.61	4.36	6.35	146	2.26	1.99	0.37	0	0.79	0.42

Note: Z value represents the Z-statistic of the K-S test, p value is the corresponding probability, and $p > 0.05$ means a normal distribution. * and # respectively refer to the statistical results of the original and logarithmic soil gravimetric water contents.

The normality test is a prerequisite for spatial analysis of soil gravimetric water content by using geostatistical methods. From Table 2, the kurtosis and skewness coefficients of soil gravimetric water content at different depths show an "increase–decrease–increase" trend with the increase in soil depths. According to the Kolmogorov–Smirnov test [37], the soil gravimetric water content at different depths does not obey a normal distribution but obeys a lognormal distribution.

3.2. Spatial Heterogeneity of Soil Moisture in the Alpine Valley Desert

The structural parameters of three semi-variance function models are shown in Table 3. The optimal semi-variance function model for soil gravimetric water content in the 0–10 cm soil layer is determined as the Spherical model, and that for soil gravimetric water content in the 10–20 and 40–50 cm soil layers is selected as the Exponential model. The optimal theoretical model for soil gravimetric water content in the 20–30 and 30–40 cm soil layers is confirmed as the Gaussian model.

The nugget values of the different soil layers are between 0.1 and 0.34, the nugget value of the 10–20 cm soil layer is the smallest, and that of the 20–30 cm soil layer is the largest. This finding indicates that the spatial heterogeneity is smallest and largest in the 10–20 and 20–30 cm soil layers, respectively. The sill value of the dune soil first increases and then decreases with the increase in soil depths. The 0–30 cm layer is greatly affected by the elevation of the soil, while the 30–50 cm layer is weakened by the elevation. The different elevations affect the growth and distribution of vegetation in different soil layers of dunes, thus impacting the spatial distribution of soil moisture.

Table 3. Theoretical semi-variogram models of soil moisture and its related parameters.

Depth /cm	Theoretical Model	Nugget	Sill	Nugget Coefficient /%	Range /m	Coefficient of Determination	Residual Sum of Squares (%^{2})
0–10	Spherical model	6.30	45.99	13.7	37.42	0.991	3.98
	Exponential model	2.40	45.8	5.20	44.61	0.98	9.49
	Gaussian model	10.73	41.98	25.60	27.31	0.97	9.99
10–20	Spherical model	14.90	111.9	23.30	23.51	0.96	110
	Exponential model	0.10	124.1	0.10	31.23	0.98	43.60
	Gaussian model	25	108.60	23	17.58	0.96	110
20–30	Spherical model	20.60	169	12.20	61.09	0.96	97.70
	Exponential model	17.90	236.70	7.60	142.28	0.95	150
	Gaussian model	34	148.90	22.80	41.70	0.99	21.20
30–40	Spherical model	20	101	19.80	45.99	0.92	142
	Exponential model	17.60	96.20	18.30	71.01	0.87	180
	Gaussian model	28.50	108	26.40	46.19	0.97	50.60
40–50	Spherical model	9.37	40.78	32	23.76	0.983	44.20
	Exponential model	7.20	53.64	13.40	63.38	0.99	0.37
	Gaussian model	13.45	40.05	33.60	23.48	0.99	5.15

From Table 3, the nugget coefficients are 13.7% in the 0–10 cm, 0.1% in the 10–20 cm, 22.8% in the 20–30 cm, 26.4% in the 30–40 cm, and 13.4% in the 40–50 cm layer. This result indicates that the soil gravimetric water content in the 30–40 cm soil layer of the dune has a moderate spatial correlation, whereas the soil gravimetric water content in other soil layers has a strong spatial correlation. The spatial correlation of the 10–20 cm soil layer is stronger than those of the other soil layers. The range of soil gravimetric water content varies from 31.23 to 63.38 m, and is greater than the minimum sampling interval of this study (15 m). This finding suggests that the sample design is sufficient to reflect the entire sample soil moisture content of the spatial structure characteristics.

To describe the spatial distribution of soil gravimetric water content intuitively, the optimal semi-variance function model and its characteristic parameters in Table 3 were selected to draw spatial distribution maps of soil gravimetric water content at different levels through the Kriging interpolation, as shown in Figure 3. The spatial distribution of soil gravimetric water content in different soil layers is mostly patched and banded. The spatial distribution of soil gravimetric water content in the sampling points is higher in the southwest than in other areas. The dune elevation gradually increases from the southwest to the northeast parts. Conversely, the soil gravimetric water content declines from the bottom of the windward slope to the top of the dune. In the 0–10, 10–20, and 20–30 cm soil layers, the soil gravimetric water content presents a banded and patchy distribution, which can reflect the soil gravimetric water content heterogeneity of different parts (Figure 3).

Table 4 shows the correlation matrix of soil gravimetric water content at different layers based on Pearson correlation analysis. There is a high similarity observed among the layers of soil moisture. The soil gravimetric water content presents an extremely significant correlation among different depths ($p < 0.01$). In most cases, the correlation of soil gravimetric water content gradually decreases with the increase in soil depths.

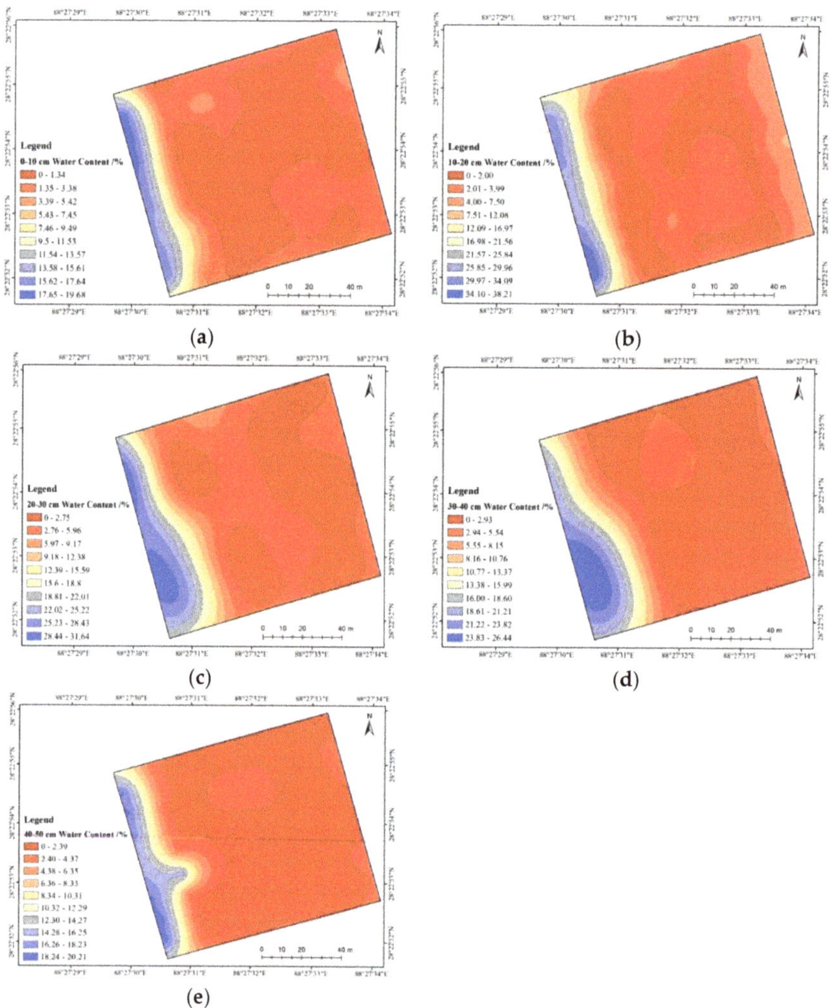

Figure 3. Spatial distribution of soil moisture in different soil layers. (**a**) soil gravimetric water content at 0–10 cm layer, (**b**) soil gravimetric water content at 10–20 cm layer, (**c**) soil gravimetric water content at 20–30 cm layer, (**d**) soil gravimetric water content at 30–40 cm layer, (**e**) soil gravimetric water content at 40–50 cm layer.

Table 4. Correlation matrix of soil moisture at different layers.

Soil Depth/cm	0–10	10–20	20–30	30–40	40–50
0–10	1				
10–20	0.953 **	1			
20–30	0.890 **	0.858 **	1		
30–40	0.820 **	0.796 **	0.970 **	1	
40–50	0.898 **	0.926 **	0.862 **	0.816 **	1

Note: ** refers to a significant correlation at the 0.01 level (bilateral).

3.3. Driving Factors of Soil Moisture in Alpine Valley Dunes

3.3.1. Factor Detector

The geographical detector can explain the influence of each factor on the change in soil moisture content in the dune. Table 5 shows the q values of evaluation factors. The ranking of the influence of the evaluation factors on the alpine valley dune is elevation > slope > location > vegetation > aspect.

Table 5. q values of evaluation factors.

Soil Depth/cm	Natural Factors	Location	Elevation	Aspect	Slope	Vegetation
0–10	q	0.620	0.881	0.001	0.620	0.478
	p value	0.000	0.430	0.859	0.000	0.000
10–20	q	0.824	0.909	0.035	0.824	0.628
	p value	0.000	0.243	0.342	0.000	0.000
20–30	q	0.737	0.928	0.028	0.737	0.682
	p value	0.000	0.137	0.222	0.000	0.000
30–40	q	0.785	0.953	0.008	0.785	0.703
	p value	0.000	0.021	0.543	0.000	0.000
40–50	q	0.738	0.928	0.000	0.738	0.592
	p value	0.000	0.109	0.926	0.000	0.000
0–50	q	0.885	0.981	0.001	0.885	0.776
	p value	0.000	0.000	0.879	0.000	0.000

The q values of elevation, slope, and vegetation coverage increase with depth. This result implies that these factors are the main influencing factors of soil gravimetric water content variation at different depths of the dune. The q values of the slope aspect do not exceed 0.05, indicating that the aspect has a minimal influence on the variability in soil gravimetric water content at different depths of the dune.

3.3.2. Interactive Detector

Table 6 indicates the evaluation factor interaction detection. The evaluation factors influence the soil gravimetric water content of the dune. An interaction was determined among the evaluation factors, but no independent factors were found. In the 0–10 cm soil layer, 60% of the evaluation factor interaction combinations show a two-factor enhancement relationship, and 40% present a nonlinear enhancement relationship (Appendix A). The most influential interactions were the dune location and elevation combination (0.8901), elevation and aspect combination (0.8901), elevation and slope combination (0.8901), and elevation and vegetation coverage combination (0.8834). The combination of slope aspect and vegetation coverage was less influential (0.5154).

In the 10–20 cm soil layer, 90% of the evaluation factor interaction combinations show a two-factor enhancement relationship, and 10% demonstrate a nonlinear enhancement relationship. The interaction influence is relatively considerable. The greatest combinations were position and elevation (0.9124), elevation and aspect (0.9124), and elevation and slope (0.9124). The slope aspect and vegetation combination was minimum (0.6880).

In the 20–30 cm soil layer, the interaction combination of evaluation factors shows a two-factor enhancement relationship. The influence of interaction is relatively remarkable. The strongest combinations were position and elevation (0.9288), elevation and aspect (0.9288), and elevation and slope (0.9288), and the weakest combination was slope aspect and vegetation (0.6880). In the 30–40 cm soil layer, the interaction combination of slope aspect and vegetation combination (0.7208) shows a nonlinear enhancement relationship, whereas the interaction combinations of other evaluation factors indicate a two-factor enhancement relationship. The influences of interaction were between 0.7891 and 0.9560.

Table 6. Evaluation factor interaction detection.

Soil Depth /cm	Index	Location	Elevation	Aspect	Slope	Vegetation
		X1	**X2**	**X3**	**X4**	**X5**
0–10	X1	0.6203				
	X2	0.8901 (Y)	0.8814			
	X3	0.6267 (Y)	0.8901 (Y)	0.0006		
	X4	0.6267 (N)	0.8901 (Y)	0.6267 (Y)	0.6203	
	X5	0.6267 (N)	0.8834 (Y)	0.5154 (Y)	0.6267 (N)	0.4778
10–20	X1	0.8235				
	X2	0.9124 (Y)	0.9085			
	X3	0.8265 (Y)	0.9124 (Y)	0.0348		
	X4	0.8265 (N)	0.9124 (Y)	0.8265 (Y)	0.8235	
	X5	0.8267 (Y)	0.9100 (Y)	0.7698 (Y)	0.8267 (Y)	0.6275
20–30	X1	0.7371				
	X2	0.9288 (Y)	0.9275			
	X3	0.7415 (Y)	0.9288 (Y)	0.0277		
	X4	0.7415 (N)	0.9288 (Y)	0.7415 (Y)	0.7371	
	X5	0.7417 (N)	0.9287 (Y)	0.6880 (Y)	0.7417 (N)	0.6821
30–40	X1	0.7855				
	X2	0.9560 (Y)	0.9535			
	X3	0.7891 (Y)	0.9560 (Y)	0.0075		
	X4	0.7891 (N)	0.9560 (Y)	0.7891 (Y)	0.7855	
	X5	0.7891 (N)	0.9542 (Y)	0.7208 (Y)	0.7891 (N)	0.7026
40–50	X1	0.7380				
	X2	0.9300 (Y)	0.9281			
	X3	0.7423 (Y)	0.9300 (Y)	0.0002		
	X4	0.7423 (N)	0.9300 (Y)	0.7423 (Y)	0.7380	
	X5	0.7426 (Y)	0.9293 (Y)	0.6210 (Y)	0.7426 (Y)	0.5919
0–50	X1	0.8853				
	X2	0.9811 (Y)	0.9807			
	X3	0.8873 (Y)	0.9811 (Y)	0.0006		
	X4	0.8873 (N)	0.9811 (Y)	0.8873 (Y)	0.8853	
	X5	0.8873 (Y)	0.9810 (Y)	0.8127 (Y)	0.8873 (Y)	0.7756

Note: Y and N refer to significant differences of detection indicators at the 95% confidence level.

In the 40–50 cm soil layer, 60% of the evaluation factor interaction combinations show a two-factor enhancement relationship, and the interaction influence was between 0.7423 and 0.9300. Among the evaluation factor interaction combinations, 40% show a nonlinear enhancement relationship, and the interaction influences were 0.6210–0.9300. In the 0–50 cm soil layer, 70% of the evaluation factor interaction combinations show a two-factor enhancement relationship, and 30% show a nonlinear enhancement relationship. The interaction among evaluation factors has a strong effect of dual factors and a nonlinear enhancement effect on soil gravimetric water content (Appendix A).

3.3.3. Ecological Detector

Ecological detection can be used to compare whether a significant difference exists between two factors on the independent variable of soil water change. In Table 6, "Y" means that a significant difference exists between the row factor and the column factor, and "N" means that no significant difference exists between them. The factor combinations in different soil layers have distinct significant effects on soil moisture changes. Among the factor combinations in the 0–10, 20–30, and 30–40 cm soil layers, 70% have significant differences in soil moisture changes, but no significant difference exists between position and slope, position and vegetation, and slope and vegetation combinations. In the 10–20, 40–50, and 0–50 cm soil layers, 90% of factor combinations have significant differences in soil gravimetric water content changes, but no significant difference exists in the positions and slope of these three soil layers.

4. Discussion

The influencing factors of soil gravimetric water content are mainly natural and human factors, but most of the Qinghai–Tibet Plateau is dominated by natural factors [38]. Among the natural factors, the primary driver of the vegetation coverage factor is soil gravimetric water content [35]. In a previous study on soil gravimetric water content, the influence of climate factors on soil moisture is mainly reflected in the difference between water infiltration and evapotranspiration caused by rainfall and solar radiation [39]. Under certain conditions, the local climate has an important influence on the spatial-temporal pattern of soil gravimetric water content, due to the differences in rainfall and temperature at different elevations, which may be an important factor affecting the soil water pattern [40]. On a small scale, soil types, topographic factors, and vegetation factors are the main factors affecting the soil water pattern; on a large scale, climatic factors, such as rainfall and evapotranspiration, are the main factors controlling the soil water pattern [24]. The impacts of soil types, vegetation factors, and topographic factors on soil water are generally consistent with depth, such that the variation in deep soil water mainly depends on soil types, vegetation, and topographic factors [2,35,41–43]. Therefore, the factors adopted in this study are mainly different dune locations, vegetation, elevation, aspect, and slope.

Many studies have also verified the special effect of vegetation cover on soil properties in the Tibetan Plateau [42,43]. Some scholars have conducted research to assess the spatial variability of soil gravimetric water content and the temporal and spatial variations in a typical alpine meadow in the Qinghai–Tibet Plateau by using a geostatistical approach. Studies on soil moisture in the Qinghai–Tibet Plateau have mainly focused on the following four aspects: (1) inversion of surface soil moisture by using remote sensing images, (2) spatial-temporal changes of soil temperature and humidity, (3) model simulation of hydrothermal characteristics, and (4) the impacts of soil moisture and physical properties. In this study, the ranking of the influence of each evaluation factor on the alpine dune was elevation > slope > location > vegetation > aspect. This result is different from that in the Loess Plateau (location on the hill slope, vegetation cover, slope, relative elevation, and sine of the aspect) [14]. This difference may be caused by harsh climate conditions and diverse natural landscapes under an average elevation of 4505 m. The results of this study can be compared with those in other arid regions. The differences in soil moisture in various regions can be clearly compared, providing research guidance for subsequent studies.

Compared with the previous studies, such as those of [35] and [44], the spatial heterogeneity of soil gravimetric water content in an alpine desert is higher than that in alpine meadows and shrubs. This condition depends mainly on the capability of soil water conservation and the sensitivity of vegetation cover to precipitation and perched aquifer. If the sample area is extremely small, it cannot contain all information to reflect the spatial heterogeneity of the entire community. If the sample plot is exceedingly large, the difference in variables among sample plots will disappear [38]. Certain scale and zoning effects occur in the selection of sample plots. This issue was also found in this study. Thus, selecting the size of sample plots and determining the direction of the plots are important in the spatial heterogeneity research of soil water.

According to the classification by coefficient of variation, the coefficient of variation less than 10% is weak variability, and that of 10–100% is moderate variability, while the coefficient of variation larger than 100% is strong variability [45]. In this study, the variation coefficients of soil gravimetric water content in different soil layers were between 130% and 169%, showing the strong variability. This may be affected by vegetation and topographic factors, as well as extreme rainfall, sand movement, strong evaporation, and groundwater table fluctuation, which is consistent with the conclusion of Li et al. [33]. Moreover, the uncertainty resulting from the random measured samples may increase such strong soil moisture variability [13,33,46]. In the geographical detector analysis, the interaction among evaluation factors had a strong effect of dual factors and a nonlinear enhancement effect on soil gravimetric water content in the 0–50 cm layer. This implies that there are no independent factors existing in the impacts of soil gravimetric water content [21].

In general, this study is of great significance to assess the soil water status in alpine desert regions. It provides basic research for artificial vegetation restoration and regeneration and soil structure analysis of different desert types in alpine cold regions. The current study applied a geographical detector to examine the influence of different factors on soil moisture and the interaction factors in a relatively small area. Owing to the harsh field conditions in the Qinghai–Tibet Plateau area, analysis of soil moisture through point distribution, drilling, and other methods requires time and effort. In the future study, remote sensing data are urgently needed to retrieve measurements of soil water, such that the spatial distribution of soil water in a large area can be verified.

5. Conclusions

Based on soil sample collection, combined with geostatistical analysis and a geographical detector method, this study detected the spatial heterogeneity and driving factors of soil gravimetric water content in typical alpine valley dunes of the Qinghai–Tibet Plateau. The results showed that the average value of soil gravimetric water content at different depths ranged from 3.68% to 7.84%. The coefficient of variation indicated that the soil moisture content in different soil layers of the sample plot shows strong variability.

The optimal theoretical models of soil gravimetric water content in 0–50 cm layers of dunes were different. The nugget coefficient shows that the soil gravimetric water content in the dune has a strong spatial correlation, and the range of the optimal theoretical model of semi-variance function was 31.23–63.38 m, which is much larger than the 15 m sampling distance. The ranking of the influence of each evaluation factor on the alpine dune was elevation > slope > location > vegetation > aspect.

The correlation analysis of soil gravimetric water content in different soil layers showed a significant correlation ($p < 0.01$). The interaction detection of factors indicated an interaction among evaluation factors, and no factors were independent of one another. In each soil layer of 0–50 cm, the interaction among evaluation factors had a strong dual effect and a nonlinear enhancement effect on soil gravimetric water content. The significance analysis of ecological detection indicated significant differences among the factor combinations of different soil layers on soil water change. In general, the analyses of spatial distribution, heterogeneity, and driving factors of soil gravimetric water content have implications for understanding the spatial heterogeneity and driving factors of soil moisture, and guiding artificial vegetation restoration and soil structure analysis of different desert types in alpine cold desert regions. The remote sensing data is urgently recommended to retrieve measurements of soil gravimetric water content in a large area for the vegetation restoration of alpine cold regions in further work.

Author Contributions: Conceptualization, Z.Z., H.Y. and J.X.; methodology, H.Y., J.H. and B.Y.; formal analysis, H.Y., J.H. and B.Y.; writing—original draft preparation, Z.Z.; writing—review and editing, Y.Z., S.W. and J.X. All authors have read and agreed to the published version of the manuscript.

Funding: This research was financially supported by the Investigation on resource environment and biodiversity of typical mountainous areas in different climatic zones (2019FY101601-2), the original innovation project of the basic frontier scientific research program, Chinese Academy of Sciences (ZDBS-LY-DQC031), the National Natural Science Foundation of China (42071259), the Natural Science Foundation of Tibet Autonomous Region (XZ2019ZR G-61), Natural Science Foundation of Xinjiang Uygur Autonomous Region (2021D01E01), the Young Talent Growth Fund Project of Northwest Institute of Ecological Environment and Resources, Chinese Academy of Sciences (FEYS2019016), and the Youth Innovation Promotion Association of the Chinese Academy of Sciences (2019430).

Institutional Review Board Statement: Not applicable.

Informed Consent Statement: Not applicable.

Data Availability Statement: The data is available from the corresponding author upon reasonable request.

Conflicts of Interest: The authors declare no conflict of interest.

Appendix A

Table A1. Evaluation Factor Interaction Results.

Soil Depth /cm	Interaction (C)	Sum of q Values	Result	Influence
0–10	X1 ∩ X2 = 0.8901	1.5017	C > Max(q (X1), q (X2))	Two-factor enhancement
	X1 ∩ X3 = 0.6267	0.6209	C > Max(q (X1), q (X2))	Nonlinear enhancement
	X1 ∩ X4 = 0.6267	1.2406	C > q (X1) + q (X4)	Two-factor enhancement
	X1 ∩ X5 = 0.6267	1.0981	C > Max(q (X1), q (X4))	Two-factor enhancement
	X2 ∩ X3 = 0.8901	0.8820	C > Max(q (X1), q (X5))	Nonlinear enhancement
	X2 ∩ X4 = 0.8901	1.5017	C > q (X2) + q (X4)	Two-factor enhancement
	X2 ∩ X5 = 0.8834	1.3592	C > Max(q (X2), q (X4))	Two-factor enhancement
	X3 ∩ X4 = 0.6267	0.6209	C > Max(q (X2), q (X5))	Nonlinear enhancement
	X3 ∩ X5 = 0.5154	0.4784	C > q (X3) + q (X5)	Nonlinear enhancement
	X4 ∩ X5 = 0.6267	1.0981	C > q (X4) + q (X5)	Two-factor enhancement
10–20	X1 ∩ X2 = 0.9124	1.7320	C > Max(q (X4), q (X5))	Two-factor enhancement
	X1 ∩ X3 = 0.8265	0.8583	C > Max(q (X1), q (X2))	Two-factor enhancement
	X1 ∩ X4 = 0.8265	1.6470	C > Max(q (X1), q (X3))	Two-factor enhancement
	X1 ∩ X5 = 0.8267	1.4510	C > Max(q (X1), q (X4))	Two-factor enhancement
	X2 ∩ X3 = 0.9124	0.9433	C > Max(q (X1), q (X5))	Two-factor enhancement
	X2 ∩ X4 = 0.9124	1.7320	C > Max(q (X2), q (X3))	Two-factor enhancement
	X2 ∩ X5 = 0.9100	1.5360	C > Max(q (X2), q (X4))	Two-factor enhancement
	X3 ∩ X4 = 0.8265	0.8583	C > Max(q (X2), q (X5))	Two-factor enhancement
	X3 ∩ X5 = 0.7698	0.6623	C > Max(q (X3), q (X4))	Nonlinear enhancement
	X4 ∩ X5 = 0.8267	1.4510	C > q (X4) + q (X5)	Two-factor enhancement
20–30	X1 ∩ X2 = 0.9288	1.6646	C > Max(q (X4), q (X5))	Two-factor enhancement
	X1 ∩ X3 = 0.7415	0.7648	C > Max(q (X1), q (X2))	Two-factor enhancement
	X1 ∩ X4 = 0.7415	1.4742	C > Max(q (X1), q (X3))	Two-factor enhancement
	X1 ∩ X5 = 0.7417	1.4192	C > Max(q (X1), q (X4))	Two-factor enhancement
	X2 ∩ X3 = 0.9288	0.9552	C > Max(q (X1), q (X5))	Two-factor enhancement
	X2 ∩ X4 = 0.9288	1.6646	C > Max(q (X2), q (X3))	Two-factor enhancement
	X2 ∩ X5 = 0.9287	1.6096	C > Max(q (X2), q (X4))	Two-factor enhancement
	X3 ∩ X4 = 0.7415	0.7648	C > Max(q (X2), q (X5))	Two-factor enhancement
	X3 ∩ X5 = 0.6880	0.7098	C > Max(q (X3), q (X4))	Two-factor enhancement
	X4 ∩ X5 = 0.7417	1.4192	C > Max(q (X3), q (X5))	Two-factor enhancement
30–40	X1 ∩ X2 = 0.9560	1.7390	C > Max(q (X4), q (X5))	Two-factor enhancement
	X1 ∩ X3 = 0.7891	0.7930	C > Max(q (X1), q (X2))	Two-factor enhancement
	X1 ∩ X4 = 0.7891	1.5710	C > Max(q (X1), q (X3))	Two-factor enhancement
	X1 ∩ X5 = 0.7891	1.4881	C > Max(q (X1), q (X4))	Two-factor enhancement
	X2 ∩ X3 = 0.9560	0.9610	C > Max(q (X1), q (X5))	Two-factor enhancement
	X2 ∩ X4 = 0.9560	1.7390	C > Max(q (X2), q (X3))	Two-factor enhancement
	X2 ∩ X5 = 0.9542	1.6561	C > Max(q (X2), q (X4))	Two-factor enhancement
	X3 ∩ X4 = 0.7891	0.7930	C > Max(q (X2), q (X5))	Two-factor enhancement
	X3 ∩ X5 = 0.7208	0.7101	C > Max(q (X3), q (X4))	Nonlinear enhancement
	X4 ∩ X5 = 0.7891	1.4881	C > q (X4) + q (X5)	Two-factor enhancement
40–50	X1 ∩ X2 = 0.9300	1.6661	C > Max(q (X4), q (X5))	Two-factor enhancement
	X1 ∩ X3 = 0.7423	0.7382	C > Max(q (X1), q (X2))	Nonlinear enhancement
	X1 ∩ X4 = 0.7423	1.4760	C > q (X1) + q (X4)	Two-factor enhancement
	X1 ∩ X5 = 0.7426	1.3299	C > Max(q (X1), q (X4))	Two-factor enhancement
	X2 ∩ X3 = 0.9300	0.9283	C > Max(q (X1), q (X5))	Nonlinear enhancement
	X2 ∩ X4 = 0.9300	1.6661	C > q (X2) + q (X4)	Two-factor enhancement
	X2 ∩ X5 = 0.9293	1.5200	C > Max(q (X2), q (X4))	Two-factor enhancement
	X3 ∩ X4 = 0.7423	0.7382	C > Max(q (X2), q (X5))	Nonlinear enhancement
	X3 ∩ X5 = 0.6210	0.5921	C > q (X3) + q (X5)	Nonlinear enhancement
	X4 ∩ X5 = 0.7426	1.3299	C > q (X4) + q (X5)	Two-factor enhancement

Table A1. *Cont.*

Soil Depth /cm	Interaction (C)	Sum of q Values	Result	Influence
0–50	X1 ∩ X2 = 0.9811	1.8660	C > Max(q (X4), q (X5))	Two-factor enhancement
	X1 ∩ X3 = 0.8873	0.8859	C > Max(q (X1), q (X2))	Nonlinear enhancement
	X1 ∩ X4 = 0.8873	1.7706	C > q (X1) + q (X4)	Two-factor enhancement
	X1 ∩ X5 = 0.8873	1.6609	C > Max(q (X1), q (X4))	Two-factor enhancement
	X2 ∩ X3 = 0.9811	0.9813	C > Max(q (X1), q (X5))	Two-factor enhancement
	X2 ∩ X4 = 0.9811	1.8660	C > Max(q (X2), q (X3))	Two-factor enhancement
	X2 ∩ X5 = 0.9810	1.7563	C > Max(q (X2), q (X4))	Two-factor enhancement
	X3 ∩ X4 = 0.8873	0.8859	C > Max(q (X2), q (X5))	Nonlinear enhancement
	X3 ∩ X5 = 0.8127	0.7762	C > q (X3) + q (X5)	Nonlinear enhancement
	X4 ∩ X5 = 0.8873	1.6609	C > q (X4) + q (X5)	Two-factor enhancement

References

1. Li, X.; Shao, M.; Zhao, C.; Jia, X. Spatial variability of soil water content and related factors across the Hexi Corridor of China. *J. Arid. Land* **2019**, *11*, 123–134. [CrossRef]
2. Bi, H.; Li, X.; Liu, X.; Guo, M.; Li, J. A case study of spatial heterogeneity of soil moisture in the Loess Plateau, western China: A geostatistical approach. *Int. J. Sediment Res.* **2009**, *24*, 63–73. [CrossRef]
3. Seneviratne, S.I.; Corti, T.; Davin, E.; Hirschi, M.; Jaeger, E.B.; Lehner, I.; Orlowsky, B.; Teuling, A. Investigating soil moisture–climate interactions in a changing climate: A review. *Earth-Sci. Rev.* **2010**, *99*, 125–161. [CrossRef]
4. McColl, K.A.; Alemohammad, S.H.; Akbar, R.; Konings, A.G.; Yueh, S.; Entekhabi, D. The global distribution and dynamics of surface soil moisture. *Nat. Geosci.* **2017**, *10*, 100–104. [CrossRef]
5. Moore, G.W.; Jones, J.A.; Bond, B.J. How soil moisture mediates the influence of transpiration on streamflow at hourly to interannual scales in a forested catchment. *Hydrol. Process.* **2011**, *25*, 3701–3710. [CrossRef]
6. Jia, X.; Zhu, Y.; Luo, Y. Soil moisture decline due to afforestation across the Loess Plateau, China. *J. Hydrol.* **2017**, *546*, 113–122. [CrossRef]
7. Venkatesh, B.; Lakshman, N.; Purandara, B.K.; Reddy, V.B. Analysis of observed soil moisture patterns under different land coversin Western Ghats, India. *J. Hydrol.* **2011**, *397*, 281–294. [CrossRef]
8. Wei, X.; Zhou, Q.; Cai, M.; Wang, Y. Effects of Vegetation Restoration on Regional Soil Moisture Content in the Humid Karst Areas—A Case Study of Southwest China. *Water* **2021**, *13*, 321. [CrossRef]
9. Baroni, G.; Ortuani, B.; Facchi, A.; Gandolfi, C. The role of vegetation and soil properties on the spatio-temporal variability of the surface soil moisture in a maize-cropped field. *J. Hydrol.* **2013**, *489*, 148–159. [CrossRef]
10. Neelam, M.; Mohanty, B. On the Radiative Transfer Model for Soil Moisture across Space, Time and Hydro-Climates. *Remote Sens.* **2020**, *12*, 2645. [CrossRef]
11. Zhang, W.; Yi, S.; Qin, Y.; Sun, Y.; Shangguan, D.; Meng, B.; Li, M.; Zhang, J. Effects of Patchiness on Surface Soil Moisture of Alpine Meadow on the Northeastern Qinghai-Tibetan Plateau: Implications for Grassland Restoration. *Remote Sens.* **2020**, *12*, 4121. [CrossRef]
12. Herbert, C.; Pablos, M.; Vall-Llossera, M.; Camps, A.; Martínez-Fernández, J. Analyzing Spatio-Temporal Factors to Estimate the Response Time between SMOS and In-Situ Soil Moisture at Different Depths. *Remote. Sens.* **2020**, *12*, 2614. [CrossRef]
13. Guo, X.; Fu, Q.; Hang, Y.; Lu, H.; Gao, F.; Si, J. Spatial Variability of Soil Moisture in Relation to Land Use Types and Topographic Features on Hillslopes in the Black Soil (Mollisols) Area of Northeast China. *Sustainability* **2020**, *12*, 3552.
14. Feng, Q.; Zhao, W.; Qiu, Y.; Zhao, M.; Zhong, L. Spatial Heterogeneity of Soil Moisture and the Scale Variability of Its Influencing Factors: A Case Study in the Loess Plateau of China. *Water* **2013**, *5*, 1226–1242.
15. Zhu, Q.; Lin, H. Influences of soil, terrain, and crop growth on soil moisture variation from transect to farm scales. *Geoderma* **2011**, *163*, 45–54. [CrossRef]
16. Lakhankar, T.; Ghedira, H.; Temimi, M.; Azar, A.E.; Khanbilvardi, R. Effect of Land Cover Heterogeneity on Soil Moisture Retrieval Using Active Microwave Remote Sensing Data. *Remote. Sens.* **2009**, *1*, 80–91. [CrossRef]
17. Li, H.; Shen, W.; Zou, C.; Jiang, J.; Fu, L.; She, G. Spatio-temporal variability of soil moisture and its effect on vegetation in a desertified aeolian riparian ecotone on the Tibetan Plateau, China. *J. Hydrol.* **2013**, *479*, 215–225. [CrossRef]
18. López-Vicente, M.; Quijano, L.; Navas, A. Spatial patterns and stability of topsoil water content in a rainfed fallow cereal field and Calcisol-type soil. *Agric. Water Manag.* **2015**, *161*, 41–52. [CrossRef]
19. Delbari, M.; Afrasiab, P.; Gharabaghi, B.; Amiri, M.; Salehian, A. Spatial variability analysis and mapping of soil physical and chemical attributes in a salt-affected soil. *Arab. J. Geosci.* **2019**, *12*, 68.
20. Arriaga, J.; Rubio, F.R. A distributed parameters model for soil water content: Spatial and temporal variability analysis. *Agric. Water Manag.* **2017**, *183*, 101–106. [CrossRef]
21. Wang, J.F.; Li, X.H.; Christakos, G.; Liao, Y.L.; Zhang, T.; Gu, X.; Zheng, X.Y. Geographical detectors-based health risk assessment and its application in the neural tube defects study of the Heshun region, China. *Int. J. Geogr. Inf. Sci.* **2010**, *24*, 107–127. [CrossRef]
22. Du, J.; Yang, Z.G. *Tibet Autonomous Region County-Level Climate Regionalization*; China Meteorological Press: Beijing, China, 2013; pp. 73–74.

23. Prokopovich, N.P.; Bara, J.P. Soil classification system, unified. In *Applied Geology*; Finkl, C., Ed.; Encyclopedia of Earth Sciences Series; Springer: Boston, MA, USA, 1984; Volume 3.
24. Entin, J.K.; Robock, A.; Vinnikov, K.Y.; Hollinger, S.E.; Liu, S.; Namkhai, A. Temporal and spatial scales of observed soil moisture variations in the extratropics. *J. Geophys. Res. Atmos.* **2000**, *105*, 11865–11877. [CrossRef]
25. Woebbecke, D.M. Plant species identification, size, and enumeration using machine vision techniqueson near-binary images. *Proc. SPIE* **1993**, *1836*, 208–219.
26. Ren, J.; Bai, Y.C.; Wang, J.D. An efficient method for extracting vegetation coverage from digital photographs. *Remote Sens. Technol. Appl.* **2010**, *25*, 719–724.
27. Journel, A.G. *Mining Geostatistics*; Academic Press: New York, NY, USA, 1978.
28. Li, H.; Reynolds, J.F. On Definition and Quantification of Heterogeneity. *Oikos* **1995**, *73*, 280–284. [CrossRef]
29. Maroufpoor, S.; Bozorg-Haddad, O.; Chu, X.F. Chapter 9—Geostatistics: Principles and Methods. In *Handbook of Probabilistic Models*; Samui, P., Bui, D.T., Chakraborty, S., Ravinesh, C.D., Eds.; Butterworth-Heinemann: Oxford, UK, 2020; pp. 229–242.
30. Jean-Paul, C.; Pierre, D. *Geostatistics: Modeling Spatial Uncertainty*; Jhon Wiley & Sons Inc.: New York, NY, USA, 1999; p. 695.
31. Paramasivam, C.; Venkatramanan, S. An Introduction to Various Spatial Analysis Techniques. *GIS Geostat. Tech. Groundw. Sci.* **2019**, 23–30. [CrossRef]
32. Zhang, Y.; Zhang, K.-C.; An, Z.-S.; Yu, Y.-P. Quantification of driving factors on NDVI in oasis-desert ecotone using geographical detector method. *J. Mt. Sci.* **2019**, *16*, 2615–2624. [CrossRef]
33. Li, H.; Shen, W.; Lin, N.; Yuan, L.; Sun, M.; Ji, D. Spatial variability of soil moisture on aeolian sandy land in riparian ecotone of middle reaches of yarlung zangbo river valley. *Trans. Chin. Soc. Agric. Eng.* **2012**, *28*, 150–155.
34. Liu, B.; Zhao, W.; Zeng, F. Statistical analysis of the temporal stability of soil moisture in three desert regions of northwestern China. *Env. Earth Sci.* **2013**, *70*, 2249–2262. [CrossRef]
35. Zhu, X.; Shao, M.; Liang, Y.; Tian, Z.; Wang, X.; Qu, L. Mesoscale spatial variability of soilâ water content in an alpine meadow on the northern Tibetan Plateau. *Hydrol. Process.* **2019**, *33*, 2523–2534. [CrossRef]
36. Zhang, X.; Zhao, Y. Identification of the driving factors' influences on regional energy-related carbon emissions in China based on geographical detector method. *Environ. Sci. Pollut. Res.* **2018**, *25*, 9626–9635. [CrossRef]
37. Conover, W.J.; Conover, W.J. *Practical Nonparametric Statistics*; John Wiley & Sons: Hoboken, NJ, USA, 1980.
38. Liu, L.J.; Zhang, H.; Luo, L. Spatial heterogeneity of soil water of alpine area in eastern Qinghai-tibet plateau. *J. Wuhan Univ.* **2008**, *54*, 414–420.
39. Savva, Y.; Szlavecz, K.; Carlson, D.; Gupchup, J.; Szalay, A.; Terzis, A. Spatial patterns of soil moisture under forest and grass land cover in a suburban area, in Maryland, USA. *Geoderma* **2013**, *192*, 202–210. [CrossRef]
40. Wang, Y.; Shao, M.; Liu, Z. Vertical distribution and influencing factors of soil water content within 21-m profile on the Chinese Loess Plateau. *Geoderma* **2013**, 300–310. [CrossRef]
41. Zhu, Q.; Nie, X.; Zhou, X.; Liao, K.; Li, H. Soil moisture response to rainfall at different topographic positions along a mixed land-use hillslope. *Catena* **2014**, *119*, 61–70. [CrossRef]
42. Yang, R.M.; Zhang, G.L.; Yang, F.; Zhi, J.J.; Yang, F.; Liu, F.; Zhao, Y.G.; Li, D.C. Precise estimation of soil organic carbon stocks in the northeast Tibetan Plateau. *Sci. Rep.* **2016**, *6*, 21842. [CrossRef] [PubMed]
43. Zeng, C.; Zhang, F.; Wang, Q.; Chen, Y.; Joswiak, D.R. Impact of alpine meadow degradation on soil hydraulic properties over the Qinghai-Tibetan Plateau. *J. Hydrol.* **2013**, *478*, 148–156. [CrossRef]
44. Yang, Z.; Ouyang, H.; Zhang, X.; Xu, X.; Zhou, C.; Yang, W. Spatial variability of soil moisture at typical alpine meadow and steppe sites in the Qinghai-Tibetan Plateau permafrost region. *Environ. Earth Sci.* **2010**, *63*, 477–488. [CrossRef]
45. Li, H. *A Study on Spatial Variability of Soil Properties and Influence of Vegetation on It at a Catchment of Southern Jiangsu Province in China*; Nanjing Forestry University: Nanjing, China, 2008; (In Chinese with English Abstract).
46. Yang, Y.; Huang, Y.; Zhang, Y.; Tong, X. Optimal Irrigation Mode and Spatio-Temporal Variability Characteristics of Soil Moisture Content in Different Growth Stages of Winter Wheat. *Water* **2018**, *10*, 1180. [CrossRef]

Article

Soil Moisture and Salinity Inversion Based on New Remote Sensing Index and Neural Network at a Salina-Alkaline Wetland

Jie Wang [1], Weikun Wang [1], Yuehong Hu [1], Songni Tian [1] and Dongwei Liu [1,2,*]

[1] School of Ecology and Environment, Inner Mongolia University, Hohhot 010021, China; 31815035@mail.imu.edu.cn (J.W.); 31915103@mail.imu.edu.cn (W.W.); 31915132@mail.imu.edu.cn (Y.H.); 31915039@mail.imu.edu.cn (S.T.)

[2] Inner Mongolia Key Laboratory of River and Lake Ecology, Hohhot 010021, China

* Correspondence: liudw@imu.edu.cn; Tel.: +86-471-499-1436

Abstract: In arid and semi-arid regions, soil moisture and salinity are important elements to control regional ecology and climate, vegetation growth and land function. Soil moisture and salt content are more important in arid wetlands. The Ebinur Lake wetland is an important part of the ecological barrier of Junggar Basin in Xinjiang, China. The Ebinur Lake Basin is a representative area of the arid climate and ecological degradation in central Asia. It is of great significance to study the spatial distribution of soil moisture and salinity and its causes for land and wetland ecological restoration in the Ebinur Lake Basin. Based on the field measurement and Landsat 8 satellite data, a variety of remote sensing indexes related to soil moisture and salinity were tested and compared, and the prediction models of soil moisture and salinity were established, and the accuracy of the models was assessed. Among them, the salinity indexes D1 and D2 were the latest ones that we proposed according to the research area and data. The distribution maps of soil moisture and salinity in the Ebinur Lake Basin were retrieved from remote sensing data, and the correlation analysis between soil moisture and salinity was performed. Among several soil moisture and salinity prediction indexes, the normalized moisture index NDWI had the highest correlation with soil moisture, and the salinity index D2 had the highest correlation with soil salinity, reaching 0.600 and 0.637, respectively. The accuracy of the BP neural network model for estimating soil salinity was higher than the one of other models; R^2 = 0.624, RMSE = 0.083 S/m. The effect of the cubic function prediction model for estimating soil moisture was also higher than that of the BP neural network, support vector machine and other models; R^2 = 0.538, RMSE = 0.230. The regularity of soil moisture and salinity changes seemed to be consistent, the correlation degree was 0.817, and the synchronous change degree was higher. The soil salinity in the Ebinur Lake Basin was generally low in the surrounding area, high in the middle area, high in the lake area and low in the vegetation coverage area. The soil moisture in the Ebinur Lake Basin slightly decreased outward with the Ebinur Lake as the center and was higher in the west and lower in the east. However, the spatial distribution of soil moisture had a higher mutation rate and stronger heterogeneity than that of soil salinity.

Keywords: soil moisture and salinity; multispectral remote sensing; BP neural network; salina-alkaline wetland; Ebinur Lake Basin

Citation: Wang, J.; Wang, W.; Hu, Y.; Tian, S.; Liu, D. Soil Moisture and Salinity Inversion Based on New Remote Sensing Index and Neural Network at a Salina-Alkaline Wetland. *Water* **2021**, *13*, 2762. https://doi.org/10.3390/w13192762

Academic Editors: Ying Zhao, Jianguo Zhang, Jianhua Si, Jie Xue and Zhongju Meng

Received: 16 August 2021
Accepted: 25 September 2021
Published: 6 October 2021

Publisher's Note: MDPI stays neutral with regard to jurisdictional claims in published maps and institutional affiliations.

1. Introduction

Wetlands play an important role in the global ecosystem. Wetlands are easy to degrade, and their shrinkage is more obvious in arid areas. Wetlands, as an important carrier of water resources in arid areas, have irreplaceable ecological functions. In an arid area, wetlands are very sensitive to changes in soil environmental factors. In arid and semi-arid areas, changes in soil physical and chemical properties may lead to changes in wetland ecology and climate [1]. Especially in arid and semi-arid areas, the changes in soil physical and chemical properties may lead to regional ecological and climate change. Comprehensive stress of soil moisture and salinity is one of the principal factors that restrict wetland restoration in arid and semi-arid areas [2]. For example, soil salinization exerts a strong stress on the wetland and its surrounding vegetation, which leads to blocked vegetation growth and land degradation to a certain extent, and may lead to abnormal succession of ecological community, ultimately leading to the decline of wetland ecological function and productivity [3,4]. Soil moisture is used to measure available water in the soil. Soil moisture stress can result in serious desertification, degradation of vegetation productivity and change of regional microclimate in waterless areas [5]. Therefore, it is of great significance to accurately grasp the spatial distribution and formation of soil water and salt in arid and semi-arid areas for controlling natural disasters, such as land desertification and drought, as well as for protecting and restoring wetlands in arid and semi-arid areas.

In order to obtain detailed information of soil moisture and salinity distribution, traditional measurements of soil properties require large numbers of experimental samples, long-term observation of soil characteristics and surrounding vegetation information as support, which requires a huge amount of manpower, financial resources and time [6] and is not suitable for long-term monitoring due to the influence of weather. For example, in the case of bad weather conditions (rainstorm, strong wind, etc.), the field survey work of field surveyors is greatly hindered. Therefore, remote sensing has become an ideal technology for identifying, monitoring and successfully retrieving soil moisture and salinity content with its unique advantages of macroscopic, comprehensive, fast information acquisition, short cycle and dynamic reflection of the changes on the ground [7]. Soil moisture is usually retrieved by optical remote sensing or microwave remote sensing. In optical remote sensing, the common method includes using Landsat, IKONOS or MODIS multispectral data to establish the corresponding water index, drought index or vegetation index (such as the most commonly used vegetation index—the normalized vegetation index, NDVI [8,9]) to extract soil moisture [10–12]. It also contains the use of surface temperature or thermal inertia to realize soil moisture inversion [13]. Sinha et al. studied the emission characteristics of visible and near infrared spectra of alluvial soil, red loam and black cotton soil, and found that the three soils under different conditions were negatively correlated with soil moisture [14]. At the same time, some researchers have carried out studies on soil moisture and soil multispectral characteristics, indicating that visible, mid-infrared and near-infrared lights are significantly correlated with soil moisture [15]. Comparatively, spectral data in the mid-infrared band have the highest correlation with soil moisture, indicating the best effect. Ghulam et al. proposed the modified vertical drought index (MPDI) to monitor soil water content by analyzing the spectral characteristic space of soil water and found that although MPDI had a good monitoring effect, its application scope was limited [16]. Subsequently, Zhang et al. proposed a new index, RDMI, to monitor soil moisture based on visible red light and near-infrared spectral information. After verification, they found that RDMI had a strong negative correlation with soil moisture and the monitoring effect was better than that of the MPDI index [17]. Microwave remote sensing inversion of soil moisture has a certain theoretical basis, which is that there is a definite linear relationship between the backscattering coefficient and soil moisture [18]. Commonly used models are, among others, the Oh model, Dubois model and water cloud model [19]. For remote sensing inversion of soil moisture, optical sensors have a spatial resolution superior to microwave sensors and can also provide vegetation information. The algorithm is mature, but it is restricted to regional studies due to atmospheric influence. Microwave remote sensing can penetrate clouds and fog and observe all weather events. It has a complete theory, but it is severely affected by vegetation and has poor spatial resolution, which is suitable for large-scale research. Wang et al. studied the influence of different surface roughness on soil water inversion [17]. Bindlish et al. retrieved soil moisture by constructing microwave soil water scattering model (AIEM model) and found that the correlation between AIEM model and actual soil moisture was as high as 0.95, indicating that the microwave remote sensing model could well retrieve soil moisture under certain conditions [20]. In addition, some new algorithms, such as neural network [21], support vector machine [22], have been integrated into soil moisture inversions. There is a spatial correlation between soil salinity and soil moisture, and the inversion method is similar to that of soil moisture. Currently, the commonly used method is to extract salinity indices from multi-spectral data and establish a multivariate model with soil salinity. In addition, innovative modeling techniques (such as genetic algorithm, etc.) are also applied [21]. Microwave remote sensing cannot directly measure soil salinity, but its surface reflectance has a significant correlation with soil conductivity [23]. In addition, microwave remote sensing needs to remove the influence of vegetation in a vegetation-covered area. Therefore, it is most suitable for areas with a bare ground surface.

The Ebinur Lake Basin is suffering from consequences of many severe salt dust storms and the lake is drying up gradually, which is why we chose the Ebinur Lake Basin as the study area. The purpose of this study was to: (1) find a fast index and method to calculate the soil moisture and salinity of the Ebinur Lake Basin; (2) obtain the spatial distribution law of soil moisture and salinity in the Ebinur Lake Basin; (3) explain the reason of the difference of soil moisture and salinity distribution in the Ebinur Lake Basin. In this paper, based on the field measured data, the indexes with high correlation with soil salinity and soil moisture were selected from the common vegetation index, salinity index (including our newly constructed indexes D1, D2) and water index. Prediction models that included regression function model, neural network model and support vector machine regress model were established, from which the models with high fitting accuracy and verification accuracy were selected for spatial inversion of soil characteristics using Landsat 8. The distribution rules and causes of soil water and salt in the study region were analyzed to provide a scientific basis for wetland restoration in arid areas.

2. Materials and Methods

2.1. Study Area

The Ebinur Lake is the largest saltwater lake in Xinjiang, China, situated on the southwest of the Junggar Basin in central Asia [24]. The Ebinur Lake Basin boundary used in this study was obtained from DEM hydrological analysis. The Ebinur Lake Basin is located between 44°24′ N and 45°12′ N, and between 81°56′ E and 83°51′ E (Figure 1a). The Ebinur Lake is at the lowest elevation in the basin, about 189 m. The basin belongs to the temperate continental arid climate, with annual average temperature of 7–8 °C, annual average precipitation of 90.9 mm and evaporation of 1662 mm. The evaporation is much higher than the precipitation. The Boltala River, Jinghe River, Kuitun River and Akeqisu River flow into the lake from different directions, becoming the main water source for the lake area [25], but the incomes are not sufficient, and the lake is shrinking day by day. From 2004 to 2015, the area of the Ebinur Lake in a dry and wet season decreased by 461.98 km^2 and 322.04 km^2, respectively [26], which greatly accelerated the desertification process in the surrounding areas of the basin. The speed of desertification has reached 38 square kilometers per year. The northwest part of the study area is the Alashan Pass, with 164 days of annual heavy windy days, up to 185 days at most and 55.0 m/s of maximum wind speed [27]. The Ebinur Lake has a high mineralization degree and a prosperous salt ion content. The shrinkage of the lake surface results in the exposure of a large range of dry lake bottom, a large amount of salt dust in the saltwater lake was dried up and exposed to the ground, which causes serious soil salinization. The unique topography of the Ebinur Lake Basin includes a variety of landscapes, such as rocky desert, gravel desert, desert, salt desert, swamp and salty lake. The corresponding typical zonal soil is grey desert soil, grey brown desert soil and aeolian sandy soil, and the intrazonal soil is salt (salinity) soil, meadow soil and marsh soil [28] (Figure 1c). Built on keen wind and abundant salt sources, sandstorms are very common in this area, and salt dust storms frequently erupt [21]. The Ebinur Lake Basin has been the second largest source of salt and dust storms as well as sandstorms in the world. The drought is intensifying, and the ecological environment is deteriorating.

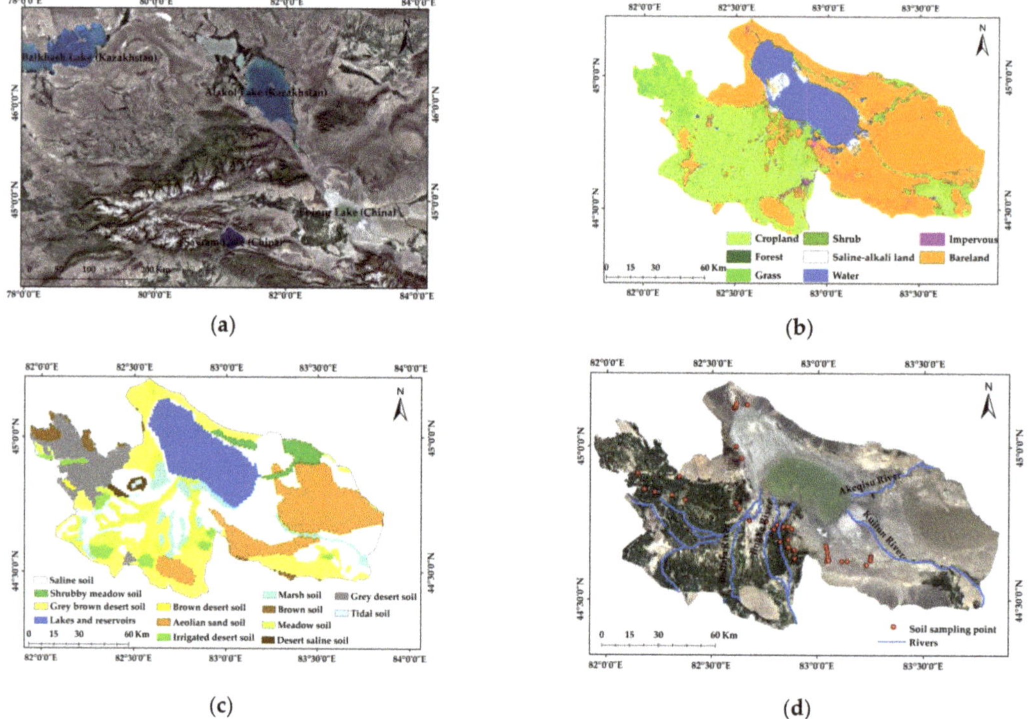

Figure 1. General map of the study area: (**a**) location of the study area; (**b**) land use in the Ebinur Lake Basin; (**c**) soil type in the Ebinur Lake Basin; (**d**) distribution map of soil sampling points.

2.2. Field Soil Sampling Experiment

Field soil experiment was made on 15 April 2017. In order to better map the soil moisture and salinity situation in the Ebinur Lake Basin, 60 sampling points were randomly distributed. The distribution of sampling points is shown in Figure 1d. At each sampling point, Stevens Portable Hydra Data Reader (Hydra-Reader) was used to obtain soil basic property data directly with Hydra probe, GPS was used to record longitude and latitude. The probe was cleaned with a polishing cloth before use to avoid increasing the error. In addition, soil specific calibration was required to ensure accuracy before measuring soil moisture [29]. At each sample point, the metal probe of the instrument was inserted vertically clockwise into the soil at a depth of about 5 cm to ensure the full contact of the soil with the metal probe. The response time was 10–20 s. The data posted by the instrument were the actual measured values, including soil type, soil temperature, soil volumetric moisture (m^3 m^{-3}), soil electrical conductivity (S/m) and temperature-corrected conductivity (S/m). Hydra-Reader has a data download service for later analysis. The electrical conductivity can be regarded as another form of expression of soil salinity. In this study, soil electrical conductivity was invoked as a measure of soil salinity, and soil volumetric moisture was used as a measure of soil moisture.

In order to ensure the accuracy of the data at the sampling points, we measured each sampling point for 6 times and took the average value as the final value. In addition, a ground object spectrometer was used to measure the reflection spectrum of the sampling point. In order to facilitate the correspondence with Landsat data, we selected the band corresponding to the central wavelength of Landsat band range in the reflection spectrum as the calculation index, so as to reduce the scale problems caused by a direct use of Landsat images. For the measured data of 60 sampling points, we randomly select 35 points as regression fitting data and 25 points as verification analysis data.

2.3. Satellite Image and Data Processing

The Landsat 8 OLI has become the data source of this study because of its convenient acquisition, moderate spatial resolution and rich band information. Its spatial resolution is 30 m, and it is free to access, including 9 bands of information, which are coastal, blue, green, red, near infrared, two short-wave infrared, two thermal infrared, panchromatic (15 m resolution) and cirrus band [30]. The date of image acquisition was 19 April 2017, which was close to the field experiment time, reducing the time error and providing more accurate soil water and salt information. The datum was obtained from the official website of the U.S. Geological Survey (USGS, https://earthexplorer.usgs.gov/ accessed on 30 July 2019).

Landsat 8 OLI data were pre-processed by the ENVI 5.3 software, including radiation calibration, atmospheric correction and mask. The ENVI 5.3 FLAASH module was used for atmospheric correction.

2.3.1. Calculation of Spectral Index

For this study, we selected some spectral indices with a certain universality and applicability related to water and salt, used by previous research institutes, as shown in Table 1. The criterion for index selection is to select those indexes that have been successfully applied in the soil water-salt inversion based on Landsat 8 data, or indexes whose application regions are in arid and semi-arid areas. SI, SI1, SI2, SI3 (four different salinity indices), S1, S2, S3, S4, S5, S6 (six different salinity calculation indicators) and NDSI (normalized salinity index) are salinity indices calculated from remote sensing data, from which surface salinity information can be obtained [31,32]. OLI_SI (Landsat 8 OLI salinity index) is a soil salinity index in an irrigation region based on Landsat 8 data. It has a strong correlation with conductivity EC [33]. SI_{vir} (visible light remote sensing salinity index) is a newly developed spectral salinity index, which has the potential to reveal soil salinity in arid climatic conditions. Int1 and Int2 are intensity indices in the visible light range, BI is a brightness index [34], which can highlight and detect different soil salinity levels [35]. NDWI (normalized water index) and LSWI (land surface water index) are water body indices, which can be used to identify land surface water body and to distinguish moisture in soil. VCI is a vegetation status index, which uses crop growth changes to reflect the degree of water information threatening crops in the region, so that it can express the regional soil moisture difference [36]. VSDI (visible and shortwave infrared drought index) is a surface drought index, which is suitable for different land cover types and less affected by vegetation. It can reflect the difference of surface soil moisture from another perspective [37]. ATI is an apparent thermal inertia, which can reflect the thermal characteristics of soil by the difference of surface temperature. It is closely related to soil moisture and has a useful application in bare soil or low vegetation coverage area [38]. EVI (enhanced vegetation index), NDVI (normalized vegetation index), SAVI (soil-adjusted vegetation index) and OSAVI (optimized soil-adjusted vegetation index) are widely used vegetation indices, which can reflect the growth status of vegetation and soil characteristics.

D1 and D2 are the spectral indices newly created for the soil salinity inversion in arid regions based on the Landsat 8 spectral characteristics of the arid region. We analyze the correlation between various bands of Landsat 8 data and the measured data, select several bands with high correlation, and combine them together. The difference between D1 and D2 is that the weight coefficients of each band is different. The weight coefficients are two sets of data obtained by analyzing the regional environment of the Ebinur Lake Basin.

The above spectral indices were calculated in the Band Math module of ENVI5.3 software.

Table 1. Spectral indices and their calculation formulas.

Index	Formula	Reference	Index	Formula	Reference
BI	$\sqrt{R_{Red}^2 + R_{NIR}^2}$	[31]	S6	$R_{NIR} \times \frac{R_{Red}}{R_{Green}}$	[31]
EVI	$2.5 \times \frac{R_{NIR} - R_{Red}}{R_{NIR} + 6 \times R_{Red} - 7.5 \times R_{Blue} + 1}$	[34]	SI	$\sqrt{R_{Blue} + R_{Red}}$	[32]
SAVI	$1.5 \times \frac{R_{NIR} - R_{Red}}{R_{NIR} + R_{Red} + 0.5}$	[39]	SI1	$\sqrt{R_{Green} \times R_{Red}}$	[40]
Int1	$(R_{Green} + R_{Red})/2$	[34]	SI2	$\sqrt{R_{Green}^2 + R_{Red}^2 + R_{NIR}^2}$	[40]
Int2	$(R_{Green} + R_{Red} + R_{NIR})/2$	[34]	SI3	$\sqrt{R_{Green}^2 + R_{Red}^2}$	[40]
OLI_SI	$50 \times R_{Coastal}^2 - R_{Blue} - R_{Green} - R_{Red}$	[33]	NDWI	$\frac{R_{Green} - R_{NIR}}{R_{Green} + R_{NIR}}$	[11]
OSAVI	$\frac{R_{NIR} - R_{Red}}{R_{NIR} + R_{Red} + 0.6}$	[41]	LSWI	$\frac{R_{NIR} - R_{SWIR}}{R_{NIR} + R_{SWIR}}$	[42]
S1	R_{Blue}/R_{Red}	[31]	VCI	$\frac{NDVI_i - NDVI_{min}}{NDVI_{max} - NDVI_{min}}$	[43]
S2	$\frac{R_{Blue} - R_{Red}}{R_{Blue} + R_{Red}}$	[31]	VSDI	$1 - \frac{R_{SWIR} - R_{Blue}}{R_{Red} - R_{Blue}}$	[37]
S3	$R_{Green} \times \frac{R_{Red}}{R_{Blue}}$	[31]	ATI	$\frac{1-\alpha}{T_d - T_n}$	[38]
S4	$\sqrt{R_{Blue} \times R_{Red}}$	[31]	SI_{vir}	$2 \times R_{Green} - R_{Red} - R_{NIR}$	[35]
S5	$R_{Blue} \times \frac{R_{Red}}{R_{Green}}$	[31]			
D1	$-8.5 \times R_{Coastal} + 15 \times R_{Blue} - 6 \times R_{Green}$				
D2	$-7 \times R_{Coastal} + 12 \times R_{Blue} - 2.5 \times R_{Green} - 1.5 \times R_{Red}$				

Note: $R_{Coastal}$, R_{Green}, R_{Blue}, R_{Red}, R_{NIR}, R_{SWIR} are the corresponding bands of Landsat 8, $NDVI_{min}$ is NDVI minimum, $NDVI_{max}$ is NDVI maximum, the pixel-by-pixel value is $NDVI_i$, α is full-band reflectance; T_d, T_n are daytime and night temperatures on the same day, respectively.

2.3.2. Inversion of Surface Temperature

The remote sensing inversion method of land surface temperature adopted in this study was the radiation transfer equation method. Its basic principle is to divide the radiation received by the sensor into three parts: surface, up-going and down-going radiation. If the intensity of atmospheric radiation is estimated, the surface radiation intensity can be obtained, and using it, the surface temperature can be calculated. The calculation formula is the Equations (1)–(3) [41].

$$L = [\varepsilon B(T_s) + (1 - \varepsilon) L_{down}]\tau + L_{up} \quad (1)$$

$$B(T_s) = [L - L_{up} - \tau(1 - \varepsilon) L_{down}]/\tau\varepsilon \quad (2)$$

$$T_s = K_2 / \ln\left(\frac{K_1}{B(T_s)} + 1\right) \quad (3)$$

where L is surface specific emissivity; T_s is surface temperature, the unit is K; B(Ts) is black-body thermal radiation brightness, L_{up} and L_{down} are upward and downward atmospheric radiations, respectively; T is atmospheric transmittance in the thermal infrared band. For Landsat 8 TIRS, $K_1 = 774.89\ w/(m^2 \times \mu m \times sr)$, $K_2 = 1321.08 \times K_1$; τ, ϵ are available on the NASA website (http://atmcorr.gsfc.nasa.gov/ accessed on 30 July 2019).

2.3.3. Construction of Temperature Vegetation Drought Index (TVDI)

TVDI is based on the simplified NDVI-Ts feature space, which is a parameter to represent soil surface characteristics. They consider that there is a certain soil moisture isoline in the characterized space [44]. In NDVI-Ts space, the pixel drought index is 1 on the dry edge, representing complete water shortage; 0 on the wet edge, representing complete water stress; and 0–1 on the data points between the dry and wet edges. The greater the TVDI value, the more serious the relative drought. The TVDI calculation is presented in Equation (4).

$$TVDI = \frac{T_S - T_{Smin}}{T_{Smax} - T_{Smin}} \quad (4)$$

where T_s is the surface temperature of a pixel; T_{smin} is the lowest temperature (wet edge temperature) at the NDVI value of the pixel; and T_{smax} is the highest temperature (dry edge temperature) at the NDVI value of the pixel.

By calculating the maximum and minimum surface temperature corresponding to each NDVI range at a certain interval, the equation of dry edge and wet edge can be fitted (Equations (5) and (6)).

$$T_{smax} = a_1 \times NDVI + b_1 \tag{5}$$

$$T_{smin} = a_2 \times NDVI + b_2 \tag{6}$$

2.3.4. Spike Cap Transformation

Spike cap transform is an orthogonal transformation of a remote sensing image, projecting image information into multi-dimensional space to obtain six components that are important to vegetation, soil and water body, such as greenness, humidity and brightness, and their shapes are similar in multidimensional space. Its shape in multi-dimensional space is the same as that of a hat, so it is called spike cap transformation. Spike cap transformation has been applied to observe the relationship between soil moisture, vegetation cover and canopy conditions [45]. Humidity band can be used to extract soil moisture information. For Landsat 8 OLI images, the conversion coefficient of humidity component is shown in Equation (7).

$$Wetness = 0.1511B_2 + 0.1973B_3 + 0.3283B_4 + 0.3407B_5 - 0.7117B_6 - 0.4559B_7 \tag{7}$$

where B_2, B_3, B_4, B_5, B_6 and B_7 are Landsat 8 blue band (Blue), green band (Green), red band (Red), near infrared band (NIR), short wave infrared 1 (SWIR1) and shortwave infrared 2 (SWIR2), respectively.

2.4. BP Neural Network Model

BP neural network is a multilayer feedforward neural network. Its main characteristics are forward signal transmission and back error propagation. In forward transmission, the input signal is processed layer by layer from the input layer to the output layer. The state of neurons in each layer only affects the state of neurons in the next layer [46]. If the output layer cannot get the expected output, it will turn to back propagation, adjust the weights and thresholds of the network according to the prediction error, so that the predictive output of BP neural network keeps approaching the expected output [47]. Before BP neural network prediction, the network must be trained to have associative memory and prediction ability. The main steps are as follows: (1) network initialization; (2) hidden layer output calculation; (3) output layer output calculation; (4) error calculation; (5) threshold updating. The commonly used algorithm of BP neural network model is gradient descent, which adjusts the changes of threshold and weight of neurons along the direction of negative gradient [48].

It can be seen from the above process that the BP neural network model has strong nonlinear mapping ability and can establish a multivariate nonlinear relationship between independent variables and dependent variables, so it is widely used in remote sensing monitoring. Soil water content is a complex nonlinear coupling system, which is influenced by topography, artificial irrigation and natural environment. Only by fully considering the influence of various factors can the inversion accuracy of soil water content and salt content be improved.

Depending on the characteristics of fitting non-linear function, the BP neural network constructed in this study determined an input parameter, namely various spectral indices, an output parameter, namely soil volume moisture or temperature-corrected conductivity. The BP neural network structure was 1–3–1; that is, there were one node in the input layer, three nodes in the hidden layer and one node in the output layer. All operations were conducted in MATLAB R2015.

2.5. SVR Support Vector Machine Regress Model

Support vector machine (SVM), such as multilayer perceptron network and radial basis function network, can be employed to pattern classification and nonlinear regression [49]. The core idea of an SVM is to establish a classification hyperplane as a decision surface to maximize the isolation edge between positive and negative examples. The theoretical basis of SVR is statistical learning theory, and more precisely, SVR is an approximate realization of structural risk minimization. SVR follows the principle of structural risk minimization and shows advantage in solving small sample, non-linear and high-dimensional pattern recognition problems. Unlike traditional machine learning methods, such as artificial neural networks, which follow the principle of empirical risk minimization, SVR avoids over-fitting, poor local optimization ability, difficulty in parameter adjustment and slow convergence. In support vector machine regression, penalty parameter C and nuclear parameter γ determine the complexity, accuracy and type of the regression model.

The introduction of kernel function can greatly improve the ability of support vector machines to deal with nonlinear problems, and at the same time maintain the intrinsic linearity of support vector machines in high dimensional space. The commonly used kernel functions mainly include linear kernel function, polynomial kernel function, radial basis kernel function and sigmoid kernel function. Since different kernel function types and parameters in SVR have a great influence on the generalization ability of the model, it is necessary to study and determine the kernel function types and parameters. Gaussian radial basis kernel function (RBF kernel function) has a good generalization ability and can support nonlinear regression [50]. In this study, we selected the RBF kernel function.

The grid method is to try the combination of various penalty parameters and RBF kernel function parameters in a certain range, and then conduct data training for each parameter combination and select the parameter combination with the best effect as the optimal parameter [51]. During data training, the k-fold cross-validation method was adopted. N-1 data were set as training data, and the rest were set as test data. The generalization error was determined by the mean value of MSE (root mean square error) after K times of calculation. The grid method parameter optimization is a violent enumeration method; if the amount of data is very large, it is a time-consuming approach, but it is also a very safe approach.

The specific operation was implemented by running the LIBSVM toolkit in the environment of MATLAB R2015b.

2.6. Correlation Analysis Model

Correlation analysis is a statistical analysis method that studies the correlation between two or more random variables with equal statuses. It is a process of describing the closeness of the relationship between objective things and using appropriate statistical indicators. The degree of correlation between the two variables is represented by the correlation coefficient r. The value of the correlation coefficient r is between -1 and 1; it can be any value within this range. In the case of positive correlation, the r value is between 0 and 1, and the scatter plot is obliquely upward. In this case, one variable increases, and the other variable increases. When the correlation is negative, the r value is between -1 and 0, and the scatter plot is diagonally downward, at which point one variable increases and the other variable decreases. The closer the absolute value of r is to 1, the stronger the correlation between the two variables, and the closer the absolute value of r is to 0, the weaker the degree of association between the two variables.

2.7. Statistical Analysis Method

The field observation data were randomly divided into analysis data and verification data, their ratio being 2:1. The basic statistical analysis of the analysis data included correlation analysis, regression analysis, linear and non-linear model fitting accuracy analysis. Among them, Pearson correlation coefficient was used as the correlation index, and its significance test was conducted. Regression analysis refers to a statistical analysis method that determines the quantitative relationship between two or more variables. Regression analysis was performed using analytical data to establish a regression model. The model with superior fitting accuracy was validated and analyzed with validation data, and the index used was root-mean-square error.

3. Results

3.1. Quantitative Analysis Results of Soil Electrical Conductivity and Remote Sensing Index

3.1.1. Correlation between Soil Electrical Conductivity and Remote Sensing Index

The correlation between remote sensing index and soil conductivity is given in Table 2. From Table 2, it can be deduced that: (1) The correlation between B1, B2, B3, B4 in Landsat 8 satellite data and soil salinity was greater than 0.4, so we chose these three bands to build the new salinity indices D1 and D2. (2) The salinity index D2 established for this study had the highest correlation with soil conductivity, followed by OLI_SI. The indexes with correlation coefficients greater than 0.6 were also D1 and Landsat 8 band 1, and their significance was less than 0.001. (3) In Landsat 8 OLI bands, bands 1 and 2 had the highest correlation with soil conductivity, followed by bands 3, 6 and 7, which had almost no correlation. (4) The correlation between vegetation index and soil conductivity was less than 0.4, and the correlation was poor. (5) The correlation between intensity indexes Int1, Int2 and soil conductivity was 0.467 and 0.414, respectively, with good significance, but the correlation of brightness index was poor. (6) The correlation between salinity index and soil conductivity was uneven. The correlation between OLI_SI, S4, S5, D1, D2 and soil conductivity was higher than 0.5, with good significance, while the correlation between S6 was very low and not significant.

Table 2. Correlation coefficients between soil electrical conductivity and remote sensing indexes.

Index	R	Index	R	Index	R
B1	0.612 **	OSAVI	−0.358 **	S5	0.521 **
B2	0.589 **	Int1	0.467 **	S6	0.127
B3	0.507 **	Int2	0.414 **	SI	0.495 **
B4	0.425 **	BI	0.349 **	SI1	0.469 **
B5	0.257 *	NDSI	0.361 **	SI2	0.404 **
B6	0.046	OLI_SI	0.637 **	SI3	0.465 **
B7	−0.066	S1	0.303 *	SI_{vir}	0.341 **
EVI	−0.274 *	S2	0.312 *	D1	0.630 **
SAVI	−0.359 **	S3	0.330 **	D2	0.650 **
NDVI	−0.361 **	S4	0.515 **	ASI	0.391 **

Note: B1, B2, B3, B4, B5, B6 and B7 are Landsat 8 coastal band, blue band, green band, red band, near infrared band, short wave infrared 1 and shortwave infrared 2, respectively; * $p < 0.05$; ** $p < 0.01$.

3.1.2. Regression Model of Soil Electrical Conductivity

The salinity index D2 was selected as the remote sensing index with the highest correlation and significant correlation with soil electrical conductivity. The regression model with soil conductivity was established. The specific parameters of regression analysis are shown in Table 3, including model formula, fitting accuracy and verification error.

In Table 3, it can be observed that: (1) The regression model of salinity index D2 and soil conductivity had good fitting accuracy. Except for the exponential model, the fitting R^2 of other models was greater than 0.5, and the validation error RMSE was less than 0.1 S/m. (2) The BP neural network model had the best fitting accuracy—R^2 was 0.624, RMSE was 0.0830 S/m (Figure 2a, verification figure)—followed by cubic function model, where R^2 was 0.620, and RMSE was 0.0834 S/m. (3) The fitting accuracy of linear function, quadratic function and SVR model was also greater than 0.6, and RMSE was less than 0.09. (4) In addition, it can be found that although the fitting accuracy of the cubic function model and the quadratic function model was high, the regression coefficients showed the over-fitting phenomenon, and the significance was not high. Therefore, in the final model selection, these two models were excluded. We selected the BP neural network model.

Table 3. Regression model and accuracy of remote sensing index and soil conductivity.

Variable	Formula	Sig	R^2	RMSE
D2	$y = 0.008e^{13.388x}$	Sig1 = 0.003, Sig2 = 0.818	0.497	0.1213
	$y = 1.195x - 0.078$	Sig1 = 0.000, Sig2 = 0.035	0.607	0.0855
	$y = 0.179\ln(x) + 0.456$	Sig1 = 0.000, Sig2 = 0.000	0.573	0.0936
	$y = -0.796x^2 + 1.481 - 0.100$	Sig1 = 0.261, Sig2 = 0.048 Sig3 = 0.260	0.609	0.0862
	$y = -25.186x^3 + 13.449x^2 - 0.928x + 0.021$	Sig1 = 0.074, Sig2 = 0.074 Sig3 = 0.074, Sig4 = 0.062	0.620	0.0834
	$y = 3.431x^{2.077}$	Sig1 = 0.001, Sig2 = 0.092	0.503	0.0928
	BP	Sig = 0.000	0.624	0.0830
	SVR	Sig = 0.001	0.497	0.1213

Note: Sig is the significance of the regression coefficient. If its value is 0.01 < Sig < 0.05, the difference is significant, and if Sig < 0.01, the difference is extremely significant. Sig1, Sig2, Sig3, Sig4 are the significances of the first, second, third and fourth coefficient, respectively.

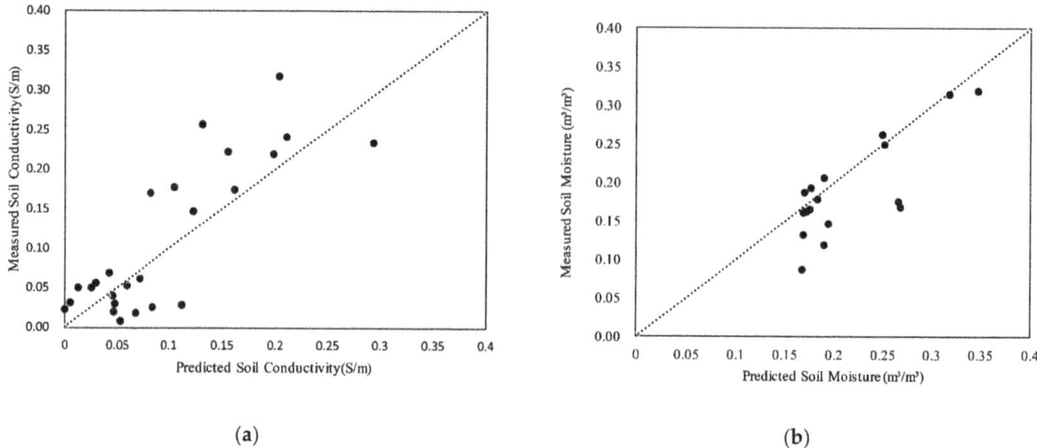

Figure 2. (**a**) Comparison of measured and predicted soil conductivity; (**b**) comparison of measured and predicted soil moisture.

3.2. *Quantitative Analysis Results of Soil Volumetric Moisture and Remote Sensing Index*

3.2.1. Correlation between Soil Volumetric Moisture and Remote Sensing Index

The correlation between remote sensing index and soil volumetric moisture is presented in Table 4. From Table 4, it can be inferred that: (1) NDWI had the highest and significant correlation with soil volumetric moisture, with R of 0.600, followed by the humidity component of the Spike cap transformation, with R of 0.572. (2) The correlation between Landsat 8 band and soil volumetric moisture was on a general level, B1 and B2 had better correlation, bands 10–11 had almost no correlation, B1–4 had positive correlation, and B5–B7 had negative correlation. (3) Vegetation indices NDVI, EVI, SAVI and VCI negatively correlated with soil volumetric moisture, with good correlations of −0.476, −0.442, −0.498 and −0.476, respectively. (4) LSWI had a low correlation with soil volumetric moisture, while TVDI, VSDI and ATI had poor and insignificant correlation. LST had almost no correlation with surface temperature.

Table 4. Coefficient of correlation between soil volumetric moisture and remote sensing indexes.

Index	R	Index	R
B1	0.468 **	EVI	−0.442 **
B2	0.435 **	SAVI	−0.498 **
B3	0.330 **	NDWI	0.600 **
B4	0.209	LSWI	0.358 **
B5	−0.063	TVDI	−0.118
B6	−0.288 *	VCI	−0.476 **
B7	−0.350 **	VSDI	0.231
B10	−0.011	ATI	0.171
B11	0.027	Wetness	0.572 **
NDVI	−0.476 **	LST	−0.006

Note: * $p < 0.05$, ** $p < 0.01$.

3.2.2. Regression Model of Soil Volumetric Moisture

Among the remote sensing indexes selected by this research institute, NDWI had the highest correlation with soil volumetric moisture. Therefore, the regression model of soil volumetric moisture was established using NDWI. The exact parameters and accuracy of the regression model are shown in Table 5. In Table 5, we can see that: (1) The fitting accuracy of the cubic function model was the best—R2 was 0.538 and RMSE was 0.230—followed by the SVR model and the quadratic function model. (2) The SVR model was not suitable as an inversion model because the fitting R^2 was high, but the verification error RMSE was the largest. (3) The fitting accuracy of exponential function was the lowest, and the fitting effect of the primary function was not good. (4) The fitting R^2 of the quadratic function was close to that of the cubic function, but the RMSE of the cubic function was smaller (Figure 2b for verification). The significance of the regression parameters was not very different. Therefore, the cubic function was chosen as the remote sensing inversion model of soil volumetric moisture.

Table 5. Regression model and accuracy of remote sensing index and soil volumetric moisture.

Variable	Formula	Sig	R2	RMSE
NDWI	$y = 1.105e^{9.185x}$	Sig1 = 0.028, Sig2 = 0.000	0.323	0.277
	$y = 4.852x + 1.163$	Sig1 = 0.001, Sig2 = 0.000	0.418	0.331
	$y = 33.002x^2 + 14.345x + 1.727$	Sig1 = 0.000, Sig2 = 0.000 Sig3 = 0.000	0.531	0.275
	$y = -105.09x^3 - 15.496x^2 + 7.906x + 1.501$	Sig1 = 0.000, Sig2 = 0.003 Sig3 = 0.021, Sig4 = 0.000	0.538	0.230
	BP	Sig = 0.001	0.502	0.277
	SVR	Sig = 0.003	0.534	0.389

3.3. Correlation Analysis of Soil Moisture and Salinity

The correlation analysis between soil electrical conductivity and volumetric moisture is shown in Figure 3. From Figure 3, it can be learned that the rules of variation of soil electrical conductivity and soil volumetric moisture tended to be the same, and the correlation degree was 0.817, which meat the synchronous change degree was higher. In other words, the correlation degree between them was higher. In addition, the change rate of soil volumetric moisture at most observation points was higher than that of soil electrical conductivity. The spatial variation rate of soil volumetric moisture was high, and its heterogeneity was strong.

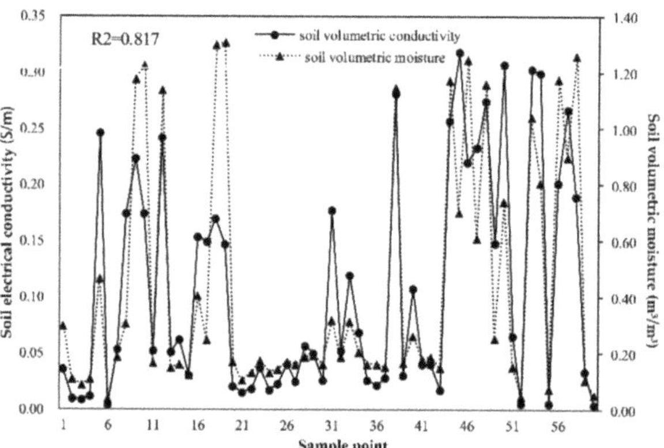

Figure 3. Simultaneous change of soil electrical conductivity and volume of soil water.

4. Discussion

4.1. Different Rules between Remote Sensing Index and Soil Moisture and Salinity

The correlation between soil electrical conductivity and Landsat 8 bands gradually decreased from B1 to B7, and there was almost no correlation between soil electrical conductivity and B6 and B7. The spectral reflectance of soil electrical conductivity increased with the increase in soil salinity. The 325–600 nm reflectance curve rose sharply with a large amount of information; the 600–1015 nm reflectance curve was flat with no obvious absorption and less information content [52]. Therefore, the correlation of Landsat 8 B1–B4 was significant, and the correlation of the B1 was the greatest. Vegetation index was negatively correlated with soil conductivity; that is, vegetation index was negatively correlated with soil salinity. Vegetation growth is susceptible to soil salinity stress, resulting in impaired photosynthesis and respiration [53]. There are only few salt-tolerant plants in the Ebinur Lake Basin, so the vegetation coverage in the areas with high soil salinity is very low. The purpose of salinity index is to highlight the information of surface salinity, so there is a high correlation between vegetation index with soil conductivity. The correlation between OLI_SI, D1, D2 and soil conductivity was greater than that of other salinity indices. These three indices were proposed for Landsat 8. The expression contained band 1 of Landsat 8, and the correlation between band 1 and soil conductivity was also higher than that of the other bands. Therefore, band 1 should be included in the subsequent study of soil salinity index for Landsat 8.

Soil volumetric moisture was positively correlated with B1–B4 of Landsat 8, and negatively correlated with B5–B7. The reason is that the reflectance spectrum of soil moisture rises rapidly at 300–750 nm with a large amount of information and tends to increase gently at 800–1350 nm with a small amount of information. There are also two steep upward slopes at 1500–1800 nm and 2100–2400 nm [38]. Water index can highlight surface moisture, with a reliable indicator effect on soil moisture. It has a significant positive correlation with soil volume moisture, especially the NDWI and humidity components of the Spike-cap transformation. In theory, the drought index and surface temperature are negatively correlated with soil moisture [12,13], but this study found that drought index and surface temperature were not correlated with soil moisture, so it is not feasible to use drought index to reflect soil moisture laterally. Drought index and surface temperature may affect soil moisture in time resolution. Generally, vegetation index is positively correlated with soil moisture [10], but in this study, vegetation index was negatively correlated with soil volumetric moisture. The reason is that the Ebinur Lake Basin is a high salinity area, where the soil contains too much salt, so that it is difficult for the vegetation to absorb water, and the growth is limited. In addition, the correction analysis of soil water and salt also validated the analysis. The change of soil water and salt was consistent, and soil salinity seriously affected the change of soil water.

4.2. Spatial Distribution of Soil Salt in Ebinur Lake Basin

The spatial distribution map of soil salinity in the Ebinur Lake Basin was obtained by model inversion, as shown in Figure 4a. From the figure, it can be seen that the soil salinity in the Ebinur Lake Basin was low around, high in the center, high in the lake area and low in the vegetation coverage area. The soil salinity in the lake area decreased gradually outward. In addition, the salt content along the coast of the Ebinur Lake, Boltala River, Jinghe River, Kuitun River and Akeqisu River is higher than that of other areas. Among them, it was the most obvious in the case of the Ebinur Lake and the Akezisu River. According to the correction analysis of soil water and salt in the Ebinur Lake Basin, the change trend of soil salinity and soil moisture tends to be the same, so in the vicinity of the water area, the soil moisture content is higher than in other areas. There are two distinct white areas in the southeastern part of the Ebinur Lake; namely, the soil electrical conductivity there is greater than 0.4 S/m. These two areas are the Jinghe Salt Field and the Jinghe Old Salt Field respectively, which further confirms the accuracy of the model inversion. The Ebinur Lake is a saltwater lake. The area of the lake is decreasing year by year. The water around the lake is gradually evaporating, and the soil salinity is gradually increasing.

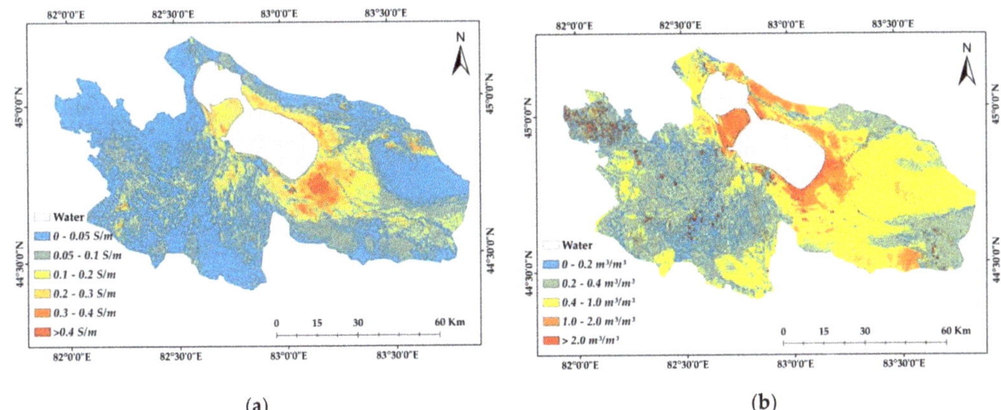

Figure 4. (**a**) Spatial distribution of soil electrical conductivity (S/m); (**b**) spatial distribution of soil volumetric moisture (m^3 m^{-3}).

From the above analysis, we can see that the soil salt content is negatively correlated with vegetation coverage. It can be learned that the soil salt content in vegetation coverage area is lower than that in bare land or wetland. In addition, soil salinity of the northern mountain forest is lower than that of the southwestern farmland. In addition to the cause of farmland fertilizer application, the high soil salinity will inhibit the growth of vegetation, so it is also suitable for planting trees and grasslands under the extremely low salinity soil. The area with soil electrical conductivity lower than 0.05 S/m in the eastern part of the country is the Arxi Sea grassland, which further validates the above discussion.

Salinization is part of the main causes of soil degradation in arid and semi-arid regions of the world. It inhibits plant growth and agricultural production and aggravates soil erosion. Nearly 20% of land in China is influenced by salinization, which is increasing with human activities, especially in arid and semi-arid areas [54]. As a typical arid and semi-arid region and an important geographic unit in central Asia, monitoring and mapping of the Ebinur Lake Basin soil salinity over a long period of time and wide space are of great significance for curbing soil degradation and sustainable agricultural production.

4.3. Spatial Distribution of Soil Moisture in Ebinur Lake Basin

The spatial distribution of soil moisture in the Ebinur Lake Basin was retrieved from the cubic regression model, as shown in Figure 4b. As can be seen on the map, soil moisture in the Ebinur Lake Basin gradually decreased outward with the Ebinur Lake as the center and was higher in the west and lower in the east. There were many white spots (i.e., soil volumetric moisture > 2 m^3 m^{-3}). In the southwest of the basin, the bright spot was the Panqiao fishpond and in the northwest, paddy fields and reservoirs. This area was the same as the lake, so the soil moisture in this area was very high. Soil volumetric moisture along the coast of the Ebinur Lake, recharged by lake water, was 1–2 m^3 m^{-3}. The soil moisture in this kind of area is also higher than that in other areas. Because of poor water storage capacity of inland saline soil, the soil moisture of inland saline soil is lower than that of other regions. The soil moisture along the four rivers is 1–2 m^3 m^{-3}, which conforms to the basic natural law.

Soil moisture content of the farmland in the west is low, and the type of soil is inland saline soil, which is not conducive to the growth of crops, but affected by topography, it developed into farmland. Under the influence of water stress, the yield of crops is low. Soil water loss is a major restrictive factor for land degradation in arid and semi-arid areas [55]. Vegetation is very vulnerable to water stress, which has a huge impact on agricultural production [56]. Therefore, timely and accurate dynamic grasp of soil moisture changes in arid and semi-arid areas is of great significance for ecological development.

4.4. Relevant Rules of Soil Water and Salt

According to the correlation analysis of soil water and salt in the Ebinur Lake Basin, the change trend of the two tended to be consistent. Depending on the physical characteristics of soil, soil water is the carrier of soil salt transport. As can be seen in Figure 4, the soil salinity was extremely high and the soil moisture was also large around the Ebinur Lake, and the two changes were related. In the eastern, southern and northwestern parts of the basin, the conductivity of soil volumetric was less than 0.05 S/m, the salinity was very low, but the soil moisture content was high. Soil moisture content was high, which has a definite dilution effect on soil salinity. Soil water and salt are related and interrelated. Especially in arid and semi-arid areas, the change of soil water and salt is one of the controlling factors in the formation of saline land [56]. It is of great significance to study the correlation and linkage effects of soil water and salt for soil restoration and inhibition of land desertification and degradation in arid areas.

4.5. Data Accuracy Discussion

The field experiment for this study was carried out in the Ebinur Lake in May, and the measurement were performed for only one year. Although multiple measurements were taken for each measurement point and their average value was taken into consideration, the inversion model of this study is not universal due to the lack of long-term continuous observation data. In arid areas, soil moisture and salt content change with the year and season. As a result, the spectral characteristics of the ground surface will change, the inversion index will also change, and the final inversion model and results will also be different. Therefore, the important direction for the future research is to study the influence of different measurement periods on the selection of soil water content and salinity inversion index, and to explore whether there is an inversion model that is more universal for all time periods.

5. Conclusions

In this study, based on a series of field experiment data of soil salinity, soil moisture and remote sensing data (from Landsat 8 OLI), the remote sensing index for estimating soil water and salt content in the Ebinur Lake Basin were tested and compared, and two new salinity indices for Landsat 8 were developed. Good models for inversion of soil moisture and salinity in the Ebinur Lake Basin were tested and obtained. The spatial distribution of soil water and salt in the Ebinur Lake Basin was predicted using the remote sensing data. The research results of this paper have certain guiding significance for the future geophysical process modeling of water and salt transport in arid saline lake basins. The basic conclusions of this study are as follows:

(1) Among the various indexes for estimating soil salinity in the Ebinur Lake Basin, the salinity index had a greater advantage than the vegetation index, and the correlation between the newly established salinity index D2 and soil conductivity was as high as 0.650. The accuracy of the BP neural network salt prediction model based on this index was also higher than other models (R^2 = 0.624, RMSE = 0.083).
(2) For the soil moisture content in the Ebinur Lake Basin, the correlation of the normalized water index (NDWI) was greater than that of other indices (R = 0.600). The cubic function prediction model had the best effect, the fitting accuracy was 0.538, and the verification error was 0.230.
(3) The correlation degree of soil water and salinity was very high, reaching 0.817. The two trends tended to be the same, but the spatial mutation rate of soil moisture was high and heterogeneity was strong.
(4) The soil salinity in the Ebinur Lake Basin was low around, high in the center, high in the lake area and low in the vegetation coverage area, and the soil salinity in the lake area decreased gradually outward. The soil moisture in the Ebinur Lake Basin gradually decreased outward with the Ebinur Lake as the center and was higher in the west and lower in the east, with more small ponds.

Author Contributions: This paper was written by J.W., D.L. reviewed and improved the manuscript with comments; the data compilation and statistical analyses were completed by J.W., W.W., Y.H., S.T., D.L. All authors have read and agreed to the published version of the manuscript.

Funding: This project is supported by Inner Mongolia major science and technology project (ZDZX2018054).

Institutional Review Board Statement: Not applicable.

Informed Consent Statement: Not applicable.

Data Availability Statement: Not applicable.

Conflicts of Interest: The authors declare no conflict of interest.

References

1. Gkiougkis, I.; Kallioras, A.; Pliakas, F.; Pechtelidis, A.; Diamantis, V.; Diamantis, I.; Ziogas, A.; Dafnis, I. Assessment of soil salinization at the eastern Nestos River Delta, N.E. Greece. *Catena* **2015**, *128*, 238–251. [CrossRef]
2. Sadeghravesh, M.H.; Khosravi, H.; Ghasemian, S. Assessment of combating-desertification strategies using the linear assignment method. *Solid Earth* **2016**, *7*, 673–683. [CrossRef]
3. Hamzeh, S.; Naseri, A.A.; AlaviPanah, S.K.; Mojaradi, B.; Bartholomeus, H.M.; Clevers, J.G.P.W.; Behzad, M. Estimating salinity stress in sugarcane fields with spaceborne hyperspectral vegetation indices. *Int. J. Appl. Earth Obs. Geoinf.* **2013**, *21*, 282–290. [CrossRef]
4. Periasamy, S.; Shanmugam, R.S. Multispectral and Microwave Remote Sensing Models to Survey Soil Moisture and Salinity. Land Degrad. Dev. **2017**, *28*, 1412–1425. [CrossRef]
5. Odunze, A.C.; Mando, A.; Sogbedji, J.; Amapu, I.Y.; Tarfa, B.D.; Yusuf, A.A.; Sunday, A.; Bello, H. Moisture conservation and fertilizer use for sustainable cotton production in the sub-humid Savanna zones of Nigeria. *Arch. Agron. Soil Sci.* **2012**, *58*, S190–S194. [CrossRef]
6. Han, L.; Liu, D.; Cheng, G.; Zhang, G.; Wang, L. Spatial distribution and genesis of salt on the saline playa at Qehan Lake, Inner Mongolia, China. *Catena* **2019**, *177*, 22–30. [CrossRef]
7. Wagner, W.; Naeimi, V.; Scipal, K.; de Jeu, R.; Martínez-Fernández, J. Soil moisture from operational meteorological satellites. *Hydrogeol. J.* **2006**, *15*, 121–131. [CrossRef]
8. Mallick, J.; AlMesfer, M.K.; Singh, V.P.; Falqi, I.I.; Singh, C.K.; Alsubih, M.; Kahla, N.B. Evaluating the NDVI–Rainfall Relationship in Bisha Watershed, Saudi Arabia Using Non-Stationary Modeling Technique. *Atmosphere* **2021**, *12*, 593. [CrossRef]
9. Kumari, N.; Srivastava, A.; Dumka, U.C. A Long-Term Spatiotemporal Analysis of Vegetation Greenness over the Himalayan Region Using Google Earth Engine. *Climate* **2021**, *9*, 109. [CrossRef]
10. Fakharizadehshirazi, E.; Sabziparvar, A.A.; Sodoudi, S. Long-term spatiotemporal variations in satellite-based soil moisture and vegetation indices over Iran. *Environ. Earth Sci.* **2019**, *78*, 342. [CrossRef]
11. Hosseini, M.; Saradjian, M.R. Multi-index-based soil moisture estimation using MODIS images. *Int. J. Remote Sens.* **2011**, *32*, 6799–6809. [CrossRef]

12. Zhu, W.; Jia, S.; Lv, A. A time domain solution of the Modified Temperature Vegetation Dryness Index (MTVDI) for continuous soil moisture monitoring. *Remote Sens. Environ.* **2017**, *200*, 1–17. [CrossRef]
13. Liu, Z.; Zhao, Y. Research on the method for retrieving soil moisture using thermal inertia model. *Sci. China Ser. D* **2006**, *49*, 539–545. [CrossRef]
14. Sinha, A. Variations in soil spectral reflectance related to soil moisture, organic matter and particle size. *J. Indian Soc. Remote Sens.* **1987**, *15*, 7–11. [CrossRef]
15. Yanjun, S.; Ying, G.; Zhaoyan, J. Monitoring soil moisture by apparent thermal intertia method. *Chin. J. Eco Agric.* **2011**, *19*, 1157–1161.
16. Ghulam, A.; Qin, Q.; Teyip, T.; Li, Z.L. Modified perpendicular drought index (MPDI): A real-time drought monitoring method. *ISPRS J. Photogramm. Remote Sens.* **2007**, *62*, 150–164. [CrossRef]
17. Zhang, J.; Zhang, Q.; Bao, A.; Wang, Y. A New Remote Sensing Dryness Index Based on the Near-Infrared and Red Spectral Space. *Remote Sens.* **2019**, *11*, 456. [CrossRef]
18. Gorrab, A.; Zribi, M.; Baghdadi, N.; Lili-Chabaane, Z.; Mougenot, B. Multi-frequency analysis of soil moisture vertical heterogeneity effect on radar backscatter. In Proceedings of the 2014 1st International Conference on Advanced Technologies for Signal and Image Processing (ATSIP), Sousse, Tunisia, 17–19 March 2014; pp. 379–384.
19. Choker, M.; Baghdadi, N.; Zribi, M.; El Hajj, M.; Paloscia, S.; Verhoest, N.E.; Lievens, H.; Mattia, F. Evaluation of the Oh, Dubois and IEM models using large dataset of SAR signal and experimental soil measurements. *Water* **2017**, *9*, 38. [CrossRef]
20. Bindlish, R.; Jackson, T.; Gasiewski, A.; Klein, M.; Njoku, E. Soil moisture mapping and AMSR-E validation using the PSR in SMEX02. *Remote Sens. Environ.* **2006**, *102*, 127–139. [CrossRef]
21. Wang, X.; Zhang, F.; Ding, J.; Kung, H.T.; Latif, A.; Johnson, V.C. Estimation of soil salt content (SSC) in the Ebinur Lake Wetland National Nature Reserve (ELWNNR), Northwest China, based on a Bootstrap-BP neural network model and optimal spectral indices. *Sci. Total Environ.* **2018**, *615*, 918–930. [CrossRef]
22. Acar, H.; Ozerdem, M.; Acar, E. Soil Moisture Inversion Via Semiempirical and Machine Learning Methods With Full-Polarization Radarsat-2 and Polarimetric Target Decomposition Data: A Comparative Study. *IEEE Access* **2020**, *8*, 197896–197907. [CrossRef]
23. Bannari, A.; Guedon, A.M.; El-Harti, A.; Cherkaoui, F.Z.; El-Ghmari, A. Characterization of Slightly and Moderately Saline and Sodic Soils in Irrigated Agricultural Land using Simulated Data of Advanced Land Imaging (EO-1) Sensor. *Commun. Soil Sci. Plant Anal.* **2008**, *39*, 2795–2811. [CrossRef]
24. Wang, J.; Ding, J.; Li, G.; Liang, J.; Yu, D.; Aishan, T.; Zhang, F.; Yang, J.; Abulimiti, A.; Liu, J. Dynamic detection of water surface area of Ebinur Lake using multi-source satellite data (Landsat and Sentinel-1A) and its responses to changing environment. *Catena* **2019**, *177*, 189–201. [CrossRef]
25. Mo, K.; Chen, Q.; Chen, C.; Zhang, J.; Wang, L.; Bao, Z. Spatiotemporal variation of correlation between vegetation cover and precipitation in an arid mountain-oasis river basin in northwest China. *J. Hydrol.* **2019**, *574*, 138–147. [CrossRef]
26. Zhu, X. Study on the Dynamic Change of Ebinur Lake Based on Multi-Source Remote Sensing Data. Master's Thesis, Xinjiang University, Xinjiang, China, 2018.
27. Zeng, H.; Wu, B.; Zhu, W.; Zhang, N. A trade-off method between environment restoration and human water consumption: A case study in Ebinur Lake. *J. Clean. Prod.* **2019**, *217*, 732–741. [CrossRef]
28. Qian, Y.; Wu, Z.; Zhang, L.; Zhou, H.; Wu, S.; Yang, Q. Eco-environmental evolution, control, and adjustment for Aibi Lake catchment. *Environ. Manag.* **2005**, *36*, 506–517. [CrossRef]
29. Vaz, C.M.P.; Jones, S.; Meding, M.; Tuller, M. Evaluation of Standard Calibration Functions for Eight Electromagnetic Soil Moisture Sensors. *Vadose Zone J.* **2013**, *12*, vzj2012.0160. [CrossRef]
30. Peng, J.; Biswas, A.; Jiang, Q.; Zhao, R.; Hu, J.; Hu, B.; Shi, Z. Estimating soil salinity from remote sensing and terrain data in southern Xinjiang Province, China. *Geoderma* **2019**, *337*, 1309–1319. [CrossRef]
31. Allbed, A.; Kumar, L.; Aldakheel, Y.Y. Assessing soil salinity using soil salinity and vegetation indices derived from IKONOS high-spatial resolution imageries: Applications in a date palm dominated region. *Geoderma* **2014**, *230–231*, 1–8. [CrossRef]
32. Ennaji, W.; Barakat, A.; Karaoui, I.; El Baghdadi, M.; Arioua, A. Remote sensing approach to assess salt-affected soils in the north-east part of Tadla plain, Morocco. *Geol. Ecol. Landsc.* **2018**, *2*, 22–28. [CrossRef]
33. El Harti, A.; Lhissou, R.; Chokmani, K.; Ouzemou, J.-E.; Hassouna, M.; Bachaoui, E.M.; El Ghmari, A. Spatiotemporal monitoring of soil salinization in irrigated Tadla Plain (Morocco) using satellite spectral indices. *Int. J. Appl. Earth Obs. Geoinf.* **2016**, *50*, 64–73. [CrossRef]
34. Bouaziz, M.; Matschullat, J.; Gloaguen, R. Improved remote sensing detection of soil salinity from a semi-arid climate in Northeast Brazil. *Comptes Rendus Geosci.* **2011**, *343*, 795–803. [CrossRef]
35. Triki Fourati, H.; Bouaziz, M.; Benzina, M.; Bouaziz, S. Modeling of soil salinity within a semi-arid region using spectral analysis. *Arab. J. Geosci.* **2015**, *8*, 11175–11182. [CrossRef]
36. Zhang, J.-H.; Zhou, Z.; Yao, F.; Yang, L.; Hao, C. Validating the Modified Perpendicular Drought Index in the North China Region Using In Situ Soil Moisture Measurement. *IEEE Geosci. Remote. Sens. Lett.* **2015**, *12*, 542–546. [CrossRef]
37. Zhang, N.; Hong, Y.; Qin, Q.; Liu, L. VSDI: A visible and shortwave infrared drought index for monitoring soil and vegetation moisture based on optical remote sensing. *Int. J. Remote Sens.* **2013**, *34*, 4585–4609. [CrossRef]
38. Song, C.; Jia, L. A Method for Downscaling FengYun-3B Soil Moisture Based on Apparent Thermal Inertia. *Remote Sens.* **2016**, *8*, 703. [CrossRef]

39. Huete, A.R. A soil-adjusted vegetation index (SAVI). *Remote Sens. Environ.* **1988**, *25*, 295–309. [CrossRef]
40. Douaoui, A.E.K.; Nicolas, H.; Walter, C. Detecting salinity hazards within a semiarid context by means of combining soil and remote-sensing data. *Geoderma* **2006**, *134*, 217–230. [CrossRef]
41. Cao, X.; Cui, X.; Yue, M.; Chen, J.; Tanikawa, H.; Ye, Y. Evaluation of wildfire propagation susceptibility in grasslands using burned areas and multivariate logistic regression. *Int. J. Remote Sens.* **2013**, *34*, 6679–6700. [CrossRef]
42. Zhang, Y.; Song, C.; Sun, G.; Band, L.E.; Noormets, A.; Zhang, Q. Understanding moisture stress on light use efficiency across terrestrial ecosystems based on global flux and remote-sensing data. *J. Geophys. Res. Biogeosci.* **2015**, *120*, 2053–2066. [CrossRef]
43. Sandholt, I.; Rasmussen, K.; Andersen, J. A simple interpretation of the surface temperature/vegetation index space for assessment of surface moisture status. *Remote Sens. Environ.* **2002**, *79*, 213–224. [CrossRef]
44. Mostafiz, C.; Chang, N.-B. Tasseled cap transformation for assessing hurricane landfall impact on a coastal watershed. *Int. J. Appl. Earth Obs. Geoinf.* **2018**, *73*, 736–745. [CrossRef]
45. Lyu, X.; Li, X.; Gong, J.; Li, S.; Huashun, D.; Dang, D.; Xuan, X.; Wang, H. Remote-sensing inversion method for aboveground biomass of typical steppe in Inner Mongolia, China. *Ecol. Indic.* **2020**, *120*, 106883. [CrossRef]
46. Hassan-Esfahani, L.; Torres-Rua, A.; Jensen, A.; McKee, M. Assessment of Surface Soil Moisture Using High-Resolution Multi-Spectral Imagery and Artificial Neural Networks. *Remote Sens.* **2015**, *7*, 2627–2646. [CrossRef]
47. Song, Q.; Wu, Y.; Soh, Y.C. Robust Adaptive Gradient-Descent Training Algorithm for Recurrent Neural Networks in Discrete Time Domain. *IEEE Trans. Neural Netw.* **2008**, *19*, 1841–1853. [CrossRef]
48. Jiang, H.; Rusuli, Y.; Amuti, T.; He, Q. Quantitative assessment of soil salinity using multi-source remote sensing data based on the support vector machine and artificial neural network. *Int. J. Remote Sens.* **2018**, *40*, 284–306. [CrossRef]
49. Hong, X.; Chen, S.; Qatawneh, A.; Daqrouq, K.; Sheikh, M.; Morfeq, A. Sparse probability density function estimation using the minimum integrated square error. *Neurocomputing* **2013**, *115*, 122–129. [CrossRef]
50. Wang, X.L.; Li, Z.B. Identifying the parameters of the kernel function in support vector machines based on the grid-search method. *Period. Ocean Univ. China* **2005**, *35*, 859–862.
51. Zhang, F. *Study on the Spectral Characteristics of Salinized Soils with Ground Objects in the Typical Oasis of Arid Area*; Xinjiang University: Xinjinag, China, 2011.
52. Fatma, M.; Asgher, M.; Masood, A.; Khan, N.A. Excess sulfur supplementation improves photosynthesis and growth in mustard under salt stress through increased production of glutathione. *Environ. Exp. Bot.* **2014**, *107*, 55–63. [CrossRef]
53. Li, J.; Pu, L.; Han, M.; Zhu, M.; Zhang, R.; Xiang, Y. Soil salinization research in China: Advances and prospects. *J. Geogr. Sci.* **2014**, *24*, 943–960. [CrossRef]
54. Ostovari, Y.; Ghorbani-Dashtaki, S.; Bahrami, H.-A.; Naderi, M.; Dematte, J.A.M. Soil loss estimation using RUSLE model, GIS and remote sensing techniques: A case study from the Dembecha Watershed, Northwestern Ethiopia. *Geoderma Reg.* **2017**, *11*, 28–36. [CrossRef]
55. Zhou, Q.; Luo, Y.; Zhou, X.; Cai, M.; Zhao, C. Response of vegetation to water balance conditions at different time scales across the karst area of southwestern China-A remote sensing approach. *Sci. Total Environ.* **2018**, *645*, 460–470. [CrossRef] [PubMed]
56. Sun, H. Dynamic Water and Salt Changes in Saline Wasteland on the Lower Edge of Plain Reservoirs in the Desert Oasis Region. *Asian Agric. Res.* **2018**, *10*, 23–33.

Article

A Characterization of the Hydrochemistry and Main Controlling Factors of Lakes in the Badain Jaran Desert, China

Bing Jia [1,2], Jianhua Si [1,*], Haiyang Xi [1] and Jie Qin [1,2]

1 Key Laboratory of Ecohydrology of Inland River Basin, Northwest Institute of Eco-Environment and Resources, Chinese Academy of Sciences (CAS)/Alxa Desert Eco-Hydrological Experimental Research Station, Lanzhou 730000, China; jiab@lzb.ac.cn (B.J.); xihy@lzb.ac.cn (H.X.); qinjie18@lzb.ac.cn (J.Q.)
2 University of Chinese Academy of Sciences, Beijing 100049, China
* Correspondence: jianhuas@lzb.ac.cn; Tel.: +86-931-496-7545

Abstract: Badain Jaran Desert, the coexistence of dunes and lakes, and the presence of the world's tallest dunes, has attracted worldwide attention among hydrologists. Freshwater, brackish, and saline lakes coexistence in the Badain Jaran Desert under extremely arid environmental conditions. This raises the question of why diverse lake water types exist under the same climatic conditions. Answering this question requires the characterization of lake hydrochemistry and the main controlling factors. The purpose of the presented research was to systematically analyzed samples from 80 lakes using statistical analysis, correlation analysis and hydrogeochemical methods to investigate the hydrochemical status and evolution of lakes in the Badain Jaran Desert. The results showed that the lake water in Badain Jaran Desert is generally alkaline, with the average pH and TDS were 9.31 and 165.12 g L^{-1}, respectively. The main cations to be Na^+ and K^+, whereas the main anions are Cl^- and SO_4^{2-}. HCO_3^- and CO_3^{2-} decreased and SO_4^{2-} and Cl^- increased from southeast to northwest, whereas lake hydrochemistry changed from the SO_4^{2-}-Cl^--HCO_3^- type to the SO_4^{2-}-Cl^- type and lakes transitioned from freshwater to saline. The freshwater and slightly brackish lakes are mainly distributed in the piedmont area at a high altitude near the Yabulai Moutains, whereas saline lakes are mainly distributed in the desert hinterland at a low altitude, and there is a roughly increasing trend of ions from the Yabulai Mountains. The evaporation-crystallization reactions are the dominant in the study region. Moreover, some saline mineral deposits, are extensive in these regions caused by intense evaporation-crystallization.

Keywords: hydrochemical characteristics; ion exchange; evaporation-crystallization; controlling factors; Badain Jaran Desert

Citation: Jia, B.; Si, J.; Xi, H.; Qin, J. A Characterization of the Hydrochemistry and Main Controlling Factors of Lakes in the Badain Jaran Desert, China. *Water* **2021**, *13*, 2931. https://doi.org/10.3390/w13202931

Academic Editor: Elias Dimitriou

Received: 13 September 2021
Accepted: 17 October 2021
Published: 19 October 2021

1. Introduction

The hydrochemistry of lake water determines its suitability for utilization in fisheries, agricultural irrigation, reed production, and salt manufacture [1]. Arid areas tend to have fragile ecological environments, characterized by scarce precipitation, intense evaporation, sparse vegetation, intense desertification, and soil erosion [2,3]. Lakes in arid regions play an important role in maintaining water resources, biodiversity, and the ecological environment. However, climate droughts and intense evaporation have resulted in the loss of water quality and quantity of lake water in desert areas, thereby reducing the potential for lake development [4].

As one of the integrated geographical units of the earth's surface, lakes are typical sites for gathering information about human activities and climate change [5]. Lake hydrochemical characteristics reflect the natural environment of the lake to some extent. Therefore, the analysis of the chemical composition of lake water is helpful to reveal water cycle processe and water-rock interactions of the leaching and evaporation of upstream rivers [6,7]. The hydrochemical characteristics of a lake can also act as an indicator of groundwater recharge and the main routes for the transport of solutes. Since the evolution of lakes in a region determines the harmony between water resources and the ecological environment, this topic has received wide attention [8–10].

The typical lakes of the arid area of the Badain Jaran Desert act to facilitate the exchange of material and energy and form an important component of the hydrological cycle [11–14]. In this region, the annual average precipitation is less than 100 mm but the evaporation rate is more than 3000 mm [15]. However, there are still over 100 permanent lakes in this extremely arid region, which vary considerably in size and hydrochemical characteristics, forming a unique megadune-lake land-scape [16,17]. In recent years, the area and quantity of lakes are decreasing. Saline marsh lakes formed by the shrinkage of lakes through evaporation are distributed in the hinterland and southeast margin of the desert [18]. These lakes, which range from fresh water to saline water, have a crucial contribution in mainating the fragile ecological balance in the desert, have become a key area of Quaternary geology [17,19–22] and lake hydrology research [7–10,23,24].

In the recent decades, the research on Badain Jaran Desert have primarily focused on the recharge of the lakes [9,11,25,26], but their hydrochemical and hydrological implications are not fully understood. As contributons from each hydrochemical source vary significantly in time and space [27], studying hydrochemistry in natural waters can provide important information about environmental change [22,28]; illustrate the complex interactions between biological, chemical, and physical subsystems in a limited area [29]; and explain the formation and evolution of water bodies using qualitative and quantitative methods [30]. Previous studies have had limited samples because of the harsh conditions, they could not exhaustively reflect the composition and distribution of the hydrochemistry of lakes in the Badain Jaran Desert. In this study, water samples from 80 lakes with different salinities in the Badain Jaran Desert were collected to comparatively study the hydrochemical characteristics and the mechanisms contributing to the hydrochemistry. Compared to the previous sampling, the present study greatly increased the catchment area and salinity range of the lakes. The objectives of the present study were to: (1) understand the hydrochemical characteristics of different types of lakes in the Badain Jaran Desert; (2) clarify the mechanisms driving evolution of different types of lakes in the Badain Jaran Desert.

2. Materials and Methods

2.1. Study Area

The Badain Jaran Desert (39.07–42.21° N, 99.39–104.57° E) is in the western part of the Inner Mongolia Autonomous Region of China, covering an area of 5.2×10^4 km^2. More specifically, the desert is east of Gurinai, south of Guaizi Lake, west of the Zongnai and Yabulai mountains, and north of Beida Mountain (Figure 1) [31]. The terrain of the desert is generally high in the south and east and low in the north and west. More specifically, elevation is higher in the Zongnai and Yabulai mountains in the east (1300–2200 m above sea level) and the Beida Mountain in the south (1800–2300 m above sea level), but gradually declines to the west and north, respectively, dropping to 970–1200 m at Guaizi Lake and east Gobi [32]. In terms of geological structure, the desert falls in the depression basin of the Alxa block. The crust has descended since the Pleistocene, thick layered Quaternary sediments have accumulated, and the surrounding mountains have strongly denuded to form a quasi-plain landform.

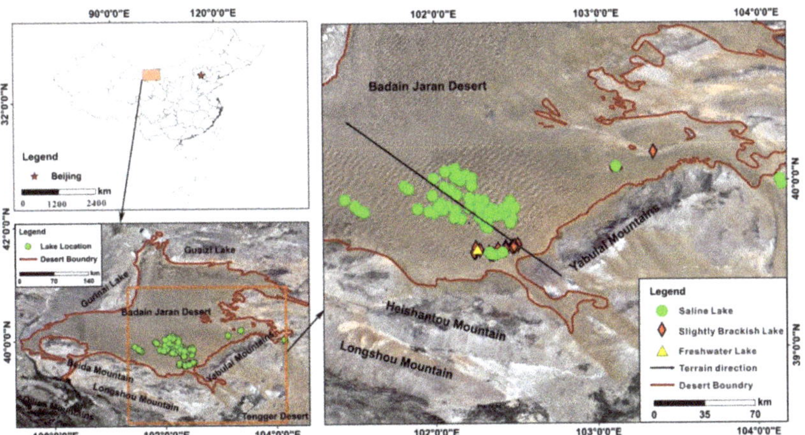

Figure 1. Map showing location of the study site and lake water samples, black line showing the terrain direction, and red lines showing the boundry of Badain Jaran Desert.The other portion indicates the sampling sites of the 80 sampled lakes in the desert with their locations highlighted as saline lake, slightly brackish lake and freshwater lake. The remote-sensing images were taken from Loca Space Viewer.

The study area belongs to the temperate continental desert steppe climate zone. The average annual rainfall decreases from 120 mm in the southeast to < 40 mm in the northwest and over half of annual rainfall falls between July and August. The potential evaporation of the lake surface can reach 1200–1550 mm [33]. There are neither rivers discharging into the lakes nor out lets draining them. The groundwater in the lake distribution area in the desert hinterland mainly occurs in the Pleistocene lacustrine sand bed. The water-resisting floor and roof comprise Tertiary argillaceous clastic rock and lacustrine sub-sandy soil and sub-clay, respectively, forming an artesian basin with a multi-layer structure. The burial depth of the water-resisting roof can be between 5–20 m and the thickness of the aquifer exceeds 30 m. The water-resisting layer in the lake area on the southern margin of the desert was formed by Cretaceous clastic rock. The aquifer is composed of Holocene lacustrine sand, sub-sand, and aeolian sand. The buried depth of the groundwater level is 1–2 m [11,34].

2.2. Study Design

In June 2019, water was sampled randomly from 80 lakes in the Badain Jaran Desert. White, high-density polyethylene (HDPE) samples bottles of volume 250 mL were sterilished in the laboratory. In the field, a multi-parameter water quality analyzer (U-53, HORIBA, JP) was used to measure water temperature, pH, total dissolved solids (TDS), salinity, conductivity (EC), redox potential (ORP), dissolved oxygen (DO), and specific gravity of seawater. A handheld global positioning system (GPSMAP 669s, GRMN, Taiwan, CHN) was used to record longitude, latitude, and elevation of sampling sites. The HDPE bottles were preinsed with the water samples three times prior to sampling. Then, each sample was collected from 0.3 to 0.5 m depth below the lake surface, filtered immediately through a 0.45 μm nylon filter, and finally stored in the precleaned HDPE bottles. The collected water samples were refrigerated at approximately 4 °C and later transported to the Laboratory of Soil and Water Chemistry, Key Laboratory of Ecohydrology of Inland River Basins, Chinese Academy of Sciences for analysis. Ca^{2+} and Mg^{2+} were determined by ethylenediaminetetraacetic acid (EDTA) titration. Na^+ and K^+ were determined by flame atomic absorption spectrophotometry. $CO_3{}^{2-}$ and $HCO_3{}^-$ were neutralized by double indicator titration. Cl^- and $SO_4{}^{2-}$ were determined by mercury nitrate titration and EDTA indirect complexation titration, respectively. The limit of detection for each ion index was 0.05 mg L^{-1}. The reliability of all water sample data were tested by cation and anion balance test, The error in the balance between cations and anions in water can be calculated with the following formula:

$$E = \frac{\sum Nc - \sum Na}{\sum Nc + \sum Na} \times 100\%, \quad (1)$$

where E represents relative error (%) and Nc and Na represent the mg equivalent (meq) concentrations of cations and anions, respectively (meq L^{-1}). The absolute value of the error in the balance between anions and cations is less than 5% for reliable data. The assessment using Equation (1) indicated that E ranged between −4.05–4.76%, indicating that data of all lake water samples were reliable.

2.3. Data Processing Method

The spatial location and topography of the water sample points were performed using ArcGIS 10.2. Statistical analysis and correlation analysis of lake hydrochemical parameters were performed using SPSS 16.0, including maximum value, minimum value, average value, standard deviation, coefficient of variation, kurtosis and skewness. A Piper plot was produced using Aqua TEChem 2014.2 to analyze the hydrochemical types of lake water. The ions interrelationships and the Gibbs diagram used to analyze the hydrochemical changes in lake were plotted using SigmaPlot 12.5 software. Visual Minteq 3.0 is used to calculate the saturation index (SI) of the main minerals in the study area.

The salinity status of lake water can be categorized according to TDS: (1) freshwater lake (TDS < 1 g L^{-1}); (2) slightly brackish lake (TDS between 1–24 g L^{-1}); (3) brackish lake (TDS between 24–35 g L^{-1}) and; (4) saline lake (TDS > 35 g L^{-1}) [35]. Figure 1 shows the distribution of sampling points from 80 lakes in the study area.

3. Results

3.1. Hydrochemical Characteristics of Lakes in the Badain Jaran Desert

The results according to TDS of water samples indicated that freshwater, slightly brackish, brackish, and saline water lakes accounted for 1.25% (one lake), 25% (20 lakes), 0.0% (zero lakes), and 73.75% (59 lakes), respectively. The pH of water samples from 80 lakes varies between 7.54 and 10.52, the average of freshwater, brackish, and saline lakes was 7.93, 8.47 ± 0.75, 9.60 ± 0.58, respectively, indicating that the lakes were overall alkaline. The TDS of freshwater, brackish, and saline lakes is 0.88 g L^{-1}, 6.76 ± 7.57 g L^{-1}, 221.54 ± 137.96 g L^{-1}, respectively. All the lakes also contained high concentrations of Na^+ (0.09–295.48 g L^{-1}) and K^+ (0.01–73.97 g L^{-1}), but the concentrations of Ca^{2+} (0.01–1.21 g L^{-1}) and Mg^{2+} (0.02–7.13 g L^{-1}) were very low. The main anions in the lake water were Cl^- (0.15–169.83 g L^{-1}) and SO_4^{2-} (0.31–110.05 g L^{-1}) and in most samples, the concentration of Cl^- was higher than that of SO_4^{2-}. Carbonate (0.02–31.37 g L^{-1}) and bicarbonate (0.15–60.33 g L^{-1}) were relatively low in the lakes.

The rankings of cations and anions ions according to average concentration were $Na^+ > K^+ > Mg^{2+} > Ca^{2+}$ and $Cl^- > SO_4^{2-} > HCO_3^- > CO_3^{2-}$, respectively. The anion and cation found in the highest concentrations in the freshwater lakes were Na^+ and SO_4^{2-}, respectively. The cations and anions found in the highest concentrations in the slightly brackish and saline lakes were Na^+ and Cl^-, respectively. High standard deviation (SD) values and differences between maximum and minimum values with regard to cations and anions indicate that the hydrochemical compositions of the lakes included in this study vary significantly. Except pH the other lake variables exhibited high variabilities in both slightly brackish and saline lakes. Kurtosis values in slightly brackish lakes are all less than zero except Mg^{2+}, while skewness values are all greater than zero except Ca^{2+}. Kurtosis values in saline lake were all greater than zero except HCO_3^-, and skewness values were all greater than zero except PH (Table 1).

Table 1. Descriptive statistics of lakes physico-chemical properties in the Badain Jaran Desert.

	Freshwater Lake	Slightly Brackish Lake							Saline Lakes						
Sample	1	20							59						
	Value	Min	Max	Mean	SD	CV	Kurtosis	Skewness	Min	Max	Mean	SD	CV	Kurtosis	Skewness
		$g\ L^{-1}$	$g\ L^{-1}$	$g\ L^{-1}$					$g\ L^{-1}$	$g\ L^{-1}$	$g\ L^{-1}$				
PH	7.93	7.65	10.1	8.47	0.75	0.09	−0.76	0.54	7.54	10.52	9.60	0.58	0.06	2.63	−1.21
TDS	0.88	1.13	20.21	6.76	7.57	1.12	−0.91	0.98	39.44	619.74	221.54	137.96	0.62	1.68	1.52
Na^+	0.09	0.30	6.53	2.08	2.45	1.17	−0.59	1.10	12.79	295.48	93.01	70.24	0.76	2.19	1.79
K^+	0.02	0.01	0.31	0.10	0.11	1.18	−0.09	1.29	0.25	73.97	8.75	14.04	1.61	9.26	2.95
Mg^{2+}	0.05	0.02	0.75	0.22	0.25	1.15	0.08	1.24	0.05	7.13	0.81	1.22	1.51	14.77	3.57
Ca^{2+}	0.07	0.01	0.07	0.04	0.02	0.47	−1.18	−0.04	0.02	1.21	0.30	0.23	0.78	3.75	1.68
Cl^-	0.15	0.22	6.94	2.10	2.55	1.21	−0.90	0.99	6.07	169.83	62.58	40.16	0.64	0.47	0.96
SO_4^{2-}	0.33	0.31	5.21	1.67	1.84	1.10	−0.25	1.17	1.33	110.05	32.52	23.67	0.73	2.29	1.46
HCO_3^-	0.25	0.15	1.86	0.66	0.56	0.85	−0.21	1.10	1.97	60.33	27.69	15.69	0.57	−0.72	0.19
CO_3^{2-}	0.06	0.02	0.67	0.23	0.23	0.99	−1.14	0.67	0.21	31.37	10.82	6.64	0.61	0.81	0.87

Min: minimum value; Max: maximum value; Mean: mean value; SD: standard deviation; CV: coefficient of variation.

The absolute content of each ion increases from freshwater lakes to saline lakes in the Badanjaran Desert. However, the relative content of ions in different lake types is varied. The meq of Na^+ and Cl^- in saline lakes were higher than that in brackish lake, and lowest in freshwater lake. Na^+ increased significantly with the transition from freshwater to saline lakes, accounting for 14%, 38%, and 52% of total ions in freshwater, slightly brackish, and saline lakes, respectively. The meq of SO_4^{2-}, HCO_3^-, Ca^{2+}, and Mg^{2+} showed decreasing trends as transition from freshwater lakes to saline lakes. The relative contents of Ca^{2+} and Mg^{2+} were higher in freshwater lakes and slightly brackish lakes, but lower in saline lakes. The meq of Ca^{2+} decreased from 12% in freshwater lakes to 3% in slightly brackish lakes and to only 0.22% in saline lakes. The meq of Mg^{2+} decreased from 13% in freshwater lakes to 9% in slightly brackish lakes and only 0.97% in saline lakes (Figure 2).

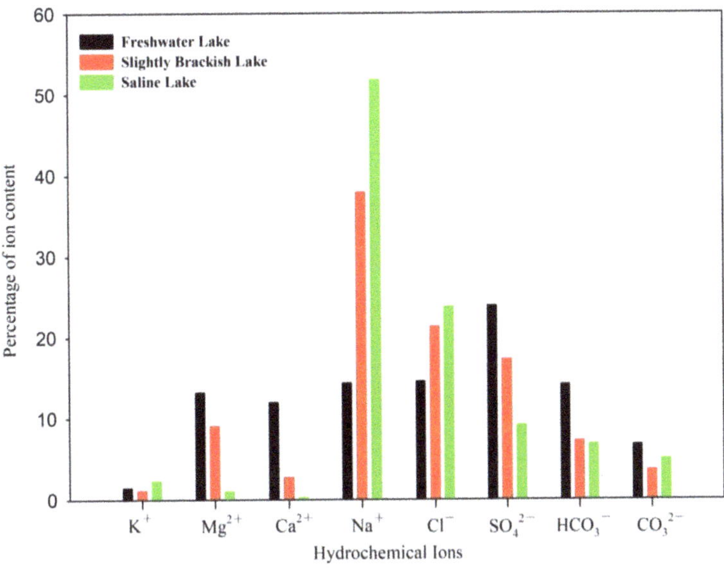

Figure 2. Ion concentrations (in meq) in individual types of lakes.

3.2. Hydrochemical Compositions and Types in the Badain Jaran Desert Lakes

The geochemistry of the water samples from the studied lakes was analyzed by plotting the concentrations of major cations and anions in the Piper-trilinear diagram (Figure 3), In the 80 studied desert lakes, except freshwater lake, the concentrations of

Na^+ and K^+ were higher than those of Ca^{2+} and Mg^{2+} except freshwater lake, hence all the points were plotted in the lower right part of the diamond. The concentration of $Cl^- + SO_4^{2-}$ in saline lake was higher than $CO_3^{2-} + HCO_3^-$, but not significant in brackish lakes. The only one sample of the freshwater lake was classified as one group, located at the middle of the diamond in the piper diagram.

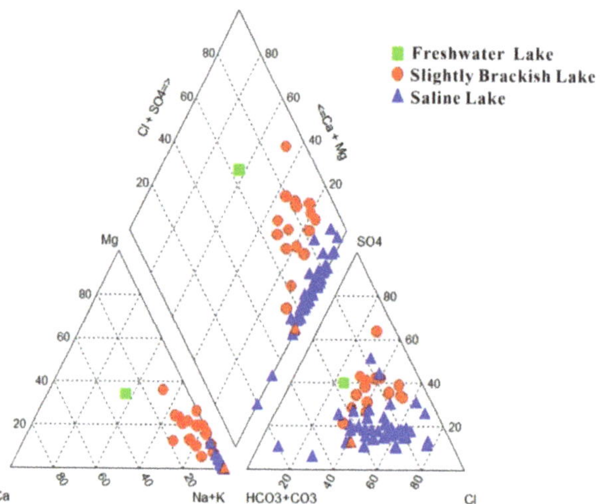

Figure 3. The Piper diagram of lake water in the Badain Jaran Desert.

The hydrochemical type in slightly brackish lake changed from Na-SO_4-Cl-HCO_3 to Na-Cl-SO_4. The hydrochemical types in saline lakes changed from the Na-Cl-HCO_3 and Na-Cl-SO_4 types to the Na-Cl type. In general, water sample contents of HCO_3^- and CO_3^{2-} decreased and SO_4^{2-} and Cl^- increased from southeast to northwest of the Badain Jaran Desert. The hydrochemistry transitioned from the SO_4-Cl-HCO_3 type to the SO_4-Cl type. As freshwater lakes transition to saline lakes, the hydrochemistry type transition from a carbonate type to the sulfate type to the chloride type.

3.3. *Mechanisms Controlling Hydrochemical Component*

3.3.1. Rock Weathering, Evaporation, and Soil-Leaching

The dominant geochemical processes of the lakes water were evaporation-crystallization, precipition dominance, and rock dominance processes in the Badain Jaran Desert (Figure 4). As shown in the Gibbs diagram that the water samples taken from freshwater and slightly brackish lakes were characterized by moderate to high ratios of $Na^+/(Na^+ + Ca^{2+})$ and $Cl^-/(Cl^- + HCO_3^-)$, indicating that evaporation-crystallization processes and rock weathering control the hydrochemical components in these regions. However, water samples from saline lakes are characterized by high concentrations of TDS and high ratios of $Na^+/(Na^+ + Ca^{2+})$ and $Cl^-/(Cl^- + HCO_3^-)$, which indicate that the water is generally dominated by products of evaporation-crystallization processes. The lake water in saline lakes evaporated more intensively than in the other lakes. The ratios of $Na^+/(Na^+ + Ca^{2+})$ in the slightly brackish lakes were more than 0.8, and that in saline lakes were close to 1.0. The ratios of $Cl^-/(Cl^- + HCO_3^-)$ in the slightly brackish lakes ranged between 0.5–0.9, and that in saline lakes ranged between 0.5–1.0.

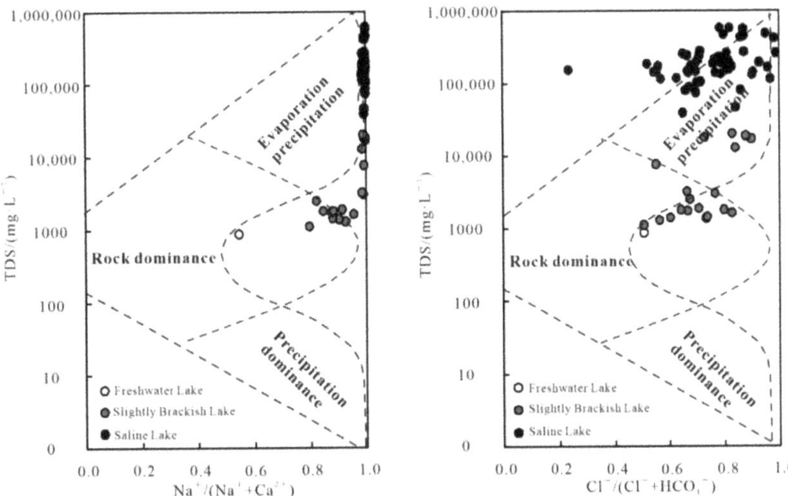

Figure 4. Gibbs diagram of the water samples in the Badain Jaran Desert.

3.3.2. Ion Exchange

Cation exchange is an important hydrogeochemical process as it modified the hydrochemical composition of groundwater and surface water. As shown in Figure 5a, the concentrations of Na^+ in most water samples from the slightly brackish lakes and saline lakes are scattered above the 1:1 line, indicating that the concentration of Na^+ in water samples exceeded that of Cl^-. SI for halite (−11.36–3.22) indicates an intensive dissolution or precipitation of secondary halite, which appears to be the cause of the high Na^+ content. Cl^- is a conserved element and is independent of other ions, Cl^- does not readily reach saturation and is relatively stable. Therefore, evaporative concentration results in Cl^- enrichment of desert lake water. The rate of Na^+ enrichment exceeded that of Cl^-, indicating strong evaporation concentration and cation exchange. Na^+ is derived not only from the multiple interactions of water in the recharge aquifer and water karst leaching, but also from cation exchange with silicate mineral Ca^{2+} (Mg^{2+}). We also compared the relationship between $(Ca^{2+} + Mg^{2+}) - (HCO_3^- + SO_4^{2-})$ and $Na^+ + K^+ - Cl^-$ (mmol L^{-1}) and found $(Ca^{2+} + Mg^{2+}) - (HCO_3^- + SO_4^{2-})$ are negatively correlated with $Na^+ + K^+ - Cl^-$ (Figure 5b). We found different correlations when the content of Na^+ in saline lakes was particularely high, so we separated seven samples from the group of hypersaline lakes and analysised them separately. The cation exchange processes are reversible, and the direction depends on the paticipating cations present in the waters and their concentrations. Therefore, we infer that the waters from the lakes have been undergoing netagive cation exchange ($Na^+ \rightarrow Ca^{2+}$, Mg^{2+}) caused by the higher concentration of Na^+. Compared with saline lake, hypersaline lake has a higher degree of nagative correlation, and silghtly brackish lake has a lower degree.

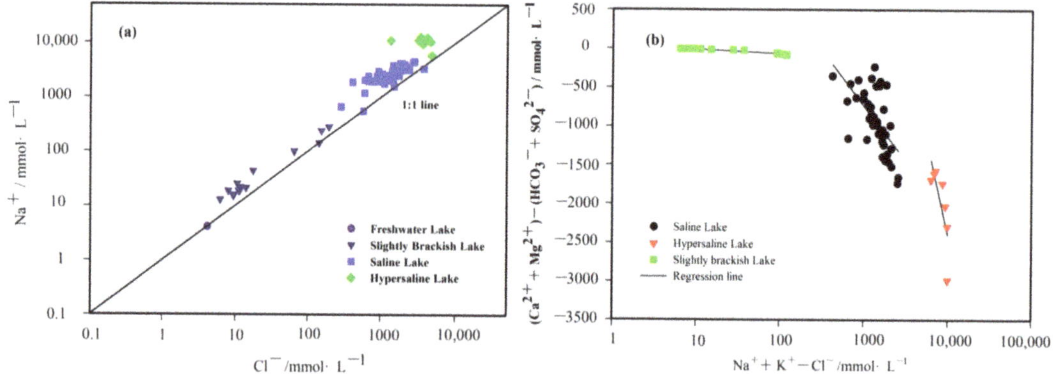

Figure 5. The relationships between (**a**) Na^+ and Cl^- and (**b**) $(Ca^{2+} + Mg^{2+}) - (HCO_3^- + SO_4^{2-})$ and $Na^+ + K^+ - Cl^-$.

3.3.3. Precipitation and Dissolution

The saturation index (SI) is an important index used to assess the degree of saturation of minerals. SI > 0 indicates deposition crystallization, whereas SI < 0 indicates mineral dissolution. Most of the SI values for anhydrite, gypsum, halite, epsomite, mirabilite, and thenardite are less than zero (Table 2), which indicates that these minerals remain undersaturated and have been dissolving. The dissolution of anhydrite, gypsum, halite, epsomite, mirabilite, and thenardite will increase the concentrations of Na^+ and SO_4^{2-} as time goes on. The dissolution of these mineral in the study area according to the following chemical reactions:

Table 2. Statistics of saturation index of main minerals in lakes water.

	Anhydrite	Aragonite	Calcite	Dolomite	Epsomite	Gypsum	Magnesite	Mirabilite	Thenardite	Halite
Mean value	−0.76	2.65	2.80	5.86	−2.92	−0.72	3.03	−1.35	−1.41	−2.53
Minimum value	−2.68	−0.35	−0.20	−0.38	−9.18	−2.48	−0.37	−6.61	−7.50	−11.36
Maximum value	2.36	6.52	6.67	13.53	−1.03	1.55	6.86	0.93	5.30	3.22

$$Na_2CO_3 + H_2O \rightarrow NaOH + NaHCO_3 \quad (2)$$

$$Na_2CO_3 + CaSO_4 \rightarrow Na_2SO_4 + CaCO_3(s) \quad (3)$$

However, most SI values for calcite, dolomite, magnesite, and aragonite are more than zero, which indicates that these minerals are supersaturated and have been precipitation. The precipitation of dolomite, calcite, and aragonite can decrease the concentrations of Ca^{2+}, Mg^{2+}, and HCO_3^-. The dissolution of these mineral in the study area according to the following chemical reactions:

$$Ca^{2+} + CO_3^{2-} \rightarrow CaCO_3(s) \quad (4)$$

$$Mg^{2+} + HCO_3^- + CaCO_3 \rightarrow CaMg(CO_3)(s) + CO_2(g) + H_2O \quad (5)$$

There was a strong linear relationship between the concentrations of $Na^+ + Cl^-$ ions and the saturation index of halite in all lakes, even more significant in hypersaline lake, indicating that the irons in water remains in an unsaturated state and that the dissolution of halite is a continuous process (Figure 6a). The relationship between Gypsum Saturation

Index and Ca^{2+} + $SO_4{}^{2-}$ is not obvious in slightly brackish lake, but linear in Saline Lake, and more obvious in hypersaline Lake (Figure 6b).

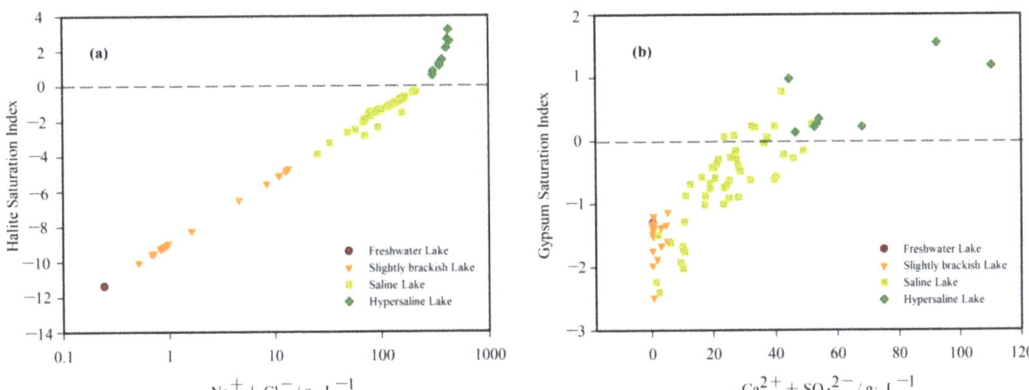

Figure 6. The relationships between (**a**) Halite Saturation Index and Na^+ + Cl^-/g L^{-1} and (**b**) Gypsum Saturation Index and Ca^{2+} + $SO_4{}^{2-}$/g L^{-1}.

4. Discussion

4.1. Variation in Hydrochemical Composition between Lakes of Different Types

The results of the present study characterized lake water of the Badain Jaran Desert as alkaline with a high TDS. This can be mainly attributed to the closed nature of lakes in the desert, with no inflowing or outflowing surface runoff and the only lake inputs being groundwater and precipitation recharge. The process of groundwater replenishment of lake water contributes to the input of not only salt, but also leaching and cation exchange with the existing environment. The process of evaporative concentration continuously enriches the salt content of the lakes, resulting in a gradual increase in TDS content and the transition of lakes to alkaline salt lakes.

In this study, we found that different types of lakes have different chemical compositions. There is little differences among the ion components in the freshwater lake. The content of cation Na^+, Ca^{2+}, Mg^{2+} were high and K^+ was low, while the content of anions shows the chararteristics of high $SO_4{}^{2-}$ and low $CO_3{}^{2-}$. The only freshwater lake in the survey, Dundejilin, had a TDS of 0.88 g/L^{-1}, lower than that measured by Shao [36] of 3.28 g/L^{-1}. The lower current TDS content of freshwater lake indicates that the lake have received large volumes of freshwater and have experienced a rapid discharge rate over an extended period of time. In addition, the lakes have been only weakly affected by rock dominance, evaporation, and crystallization.In slightly brackish lake, Na^+, Cl^- and $SO_4{}^{2-}$ were hige, but K^+, Ca^{2+} and $CO_3{}^{2-}$ were low. It is noteworthy that the coefficients of variation (CV) value for Na^+, K^+, Cl^-, and $SO_4{}^{2-}$ are very high, suggesting a large varibility of the concentration. The CV values in sligtly brackish lake were higher than that of in saline lake, indicating that hydrochemical ions are more sensitive to hydrogeological conditions, topography, hydrometeorology, and other external environmental factors in slightly brackish lake. Large coefficients of variation were noted for K^+ and Mg^{2+} in saline lakes, reflecting the strong effects of dissolution of minerals, the reaction between water and gypsum minerals, and cation exchange. The study area is characterized by low precipitation and high evaporation [37], precipitation has been considered a minor hydrochemical source compared to soil wearthing [38,39]. The ion concentrations, especially Na^+ are significantly higher than Cl^-, in saline lakes are more pronounced than in slightly brackish lakes. We therefore in fer that contributions from precipitation to lake hydrochemical compositions are lower in the saline lakes than in other lakes. As for chloride, it is more likely to dissolve from evaporates and be influenced by evaporation processes [40] because it only comes from precipitation. It is often used as a consercative reference element to study water- rock interactions because it undergoes few chemical and biological reactons in the natural environment [41]. A ratio of Na/Cl greater than 1 indicates that in addition of evaporation, minerral dissolution and cation exchange are taking place in the region.

The Badain Jaran Desert shows complex lake hydrochemical types. Most of the lakes show different hydrochemical types, even under similar hydroclimatic conditions. This phenomenon indicates a diversification of lake water sources. HCO_3^- and CO_3^{2-} in lake water decrease from southeast to northwest, whereas SO_4^{2-} and Cl^- increase, and the lake water hydrochemical type transitions from SO_4-Cl-HCO_3 to SO_4-Cl. As lakes transition from freshwater to saline lakes, their hydrochemical types transition from the carbonate type to the sulfate type to the chloride type. Water samples of different types of lakes are affected by evaporative-crystallization, and the continuous accumulation of irons results in Ca^{2+} and HCO_3^- forming insoluble irons which precipitates as well as the accumulation of soluble ions such as Na^+, K^+, and Cl^-. However, different recharge sources and recharge rates among different lakes lead to different types and rates of ion enrichment. A rapid renewal rate results in low absolute contents of Na^+ and Cl^- and a high content of Ca^{2+}, indicating an absence of sedimentation. Rapid enrichment of Na^+ and Cl^- in slightly brackish and saline lakes was noted. In addition, there was a low relative content of Ca^{2+}, indicating the occurrence of sedimentation.

4.2. Evolution of the Spatial Distribution of Hydrochemical Parameters among Different Types of Lakes

Topographic factors clearly control the distribution of slightly brackish and saline lakes, resulting in these lakes showing zonal distributions (Figure 7). Saline lakes tend to be found in relatively low-lying areas, whereas slightly brackish lakes are found in relatively high-lying areas. The larger lakes in the northern part of the desert and the lakes in the southeastern part of the desert are separated by a topographic fold and uplift, forming different groundwater recharge systems [42]. Therefore, there are significant differences in the hydrochemical types of the lakes in the region, despite their proximity. For example, the distance between Badain East Lake and Badain West Lake in the southeastern edge of the desert is less than 200 m, with the TDS of the latter lake exceeding that of the former by a factor of 100. This difference can be attributed to the higher water level of East Lake, resulting in the release of salt, whereas salt accumulates in the West Lake [43].

In our investigation, we found that the lakes on the southeast side of the desert were small and shallow, while the lakes on the north side were large and deep. The freshwater and slightly brackish lakes were mainly distributed in the piedmont area at a high altitude (1200–1300 m), whereas saline lakes were mainly distributed in the desert hinterland at a low altitude (1100–1200 m). As Figure 5 shows, Saline lakes are widely distributed in the region, while slightly brackish lakes are concentrated, mainly in the southeast of Yabuli Mountain. Among the 80 lake water samples, 21 located at the southern edge of the Badain Jaran Desert are freshwater and slightly brackish, with TDS ranging from 0.88 g L^{-1} to 20.21 g L^{-1}. 35 hyposaline lakes located in the southern-central research area show TDS values from 39.44 g L^{-1} to 199.26 g L^{-1}. Four mesosaline located in the west research area show TDS value from 279.48 g L^{-1} to 452.73 g L^{-1}. The other 10 investigated lakes are all hypersaline, located in the northwest with TDS values varying between 467.01 g L^{-1} and 619.74 g L^{-1}.Besides for Ca^{2+}, Mg^{2+} and PH, the change trend of other anions and cations is consistent with TDS. There is a roughly increasing trend of ions from the southeast to northwest, and the West Nuoertu Lake located on the northwest edge had the highest salinity while the Dundejilin Lake located on the southern edge had the lowest salinity.

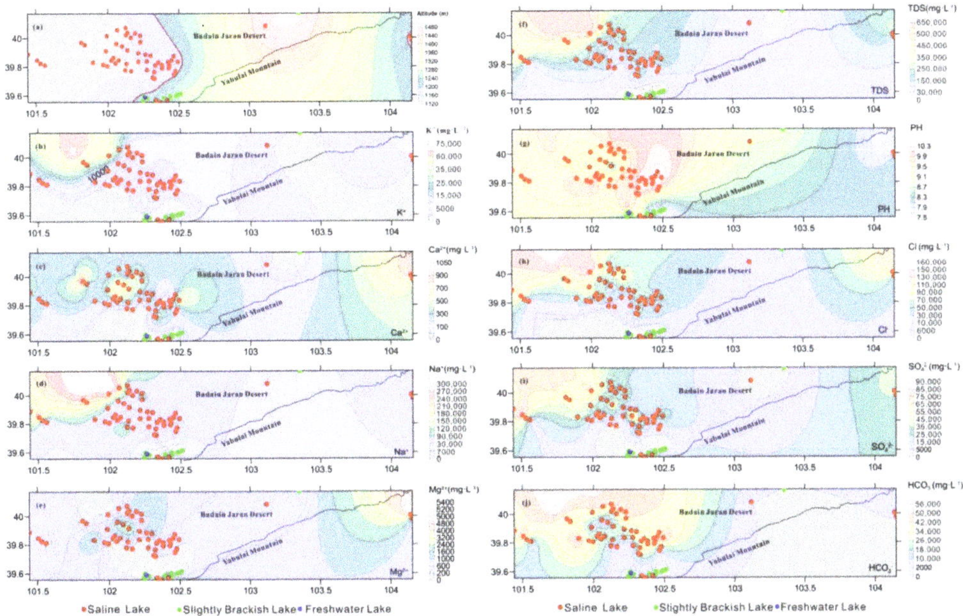

Figure 7. The spatial distribution of ion contents in the lakes of the Badain Jaran Desert.

Freshwater and slightly brackish lakes were concentrated within 30–45 km from the Yabulai Mountain. Saline lakes were widely distributed over a range of between 20–135 km. This lake distribution shown a close relationship between lake water and precipitation in Yabuli Mountain [44,45]. Due to the concentrated distribution of slightly brackish water lake, the variation trend of hydrochemical from Yabulai Mountain is not significant, we focus on the analysis of the relationship between the hydrochemical changes of saline lake and the distance from Yabulai Mountain (Figure 8). Significant changes in Cl^-, SO_4^{2-}, and Na^+ occurred. Cl^- increased from 6.07 $g \cdot L^{-1}$ to 169.83 $g \cdot L^{-1}$, SO_4^{2-} increased from 1.33 $g \cdot L^{-1}$ to 110.05 $g \cdot L^{-1}$, whereas Na^+ increased from 12.79 $g \cdot L^{-1}$ to 295.48 $g \cdot L^{-1}$. The values of TDS, Cl^-, SO_4^{2-}, HCO_3^-, CO_3^{2-} and Na^+ in most saline lakes increase with the distance from the Yabulai Mountains. Nine lakes, about 100 km from the Yabulai Mountain, have exceptionally high sodium and potassium levels. Along the terrain from the southeast to the northwest, the farther away from the Yabulai Mountains, the higher the ion content. The freshwater lakes and slightly brackish lakes in the Badain Jaran Desert are recharged by mountain precipitation and lateral runoff. The rapid rate of lake water renewal leads to a relatively low lake water TDS. The saline lakes are positioned far away from the mountain area and receive groundwater recharge. The long recharge path and complex geological conditions result in a slow lake water regeneration rate and intense evaporation, leading to relatively high TDS. This conclusion is consistent with that of Yang [46]. It is believed that the shallow groundwater is recharged by recent local precipitation infiltration near Yabulai Mountain in the southeastern margin of the desert, whereas deep groundwater is recharged by precipitation during humid periods. This assertion is consistent with the analyses by Hofmann [47] which showed the presence of two discontinuous hydraulic flow systems in this area. One is located north of the geological anticline, ~40 km from the southern edge of the desert. This flow system is recharged by precipitation. The other is located south of the geological anticline and is recharged by water from outside the dune area.

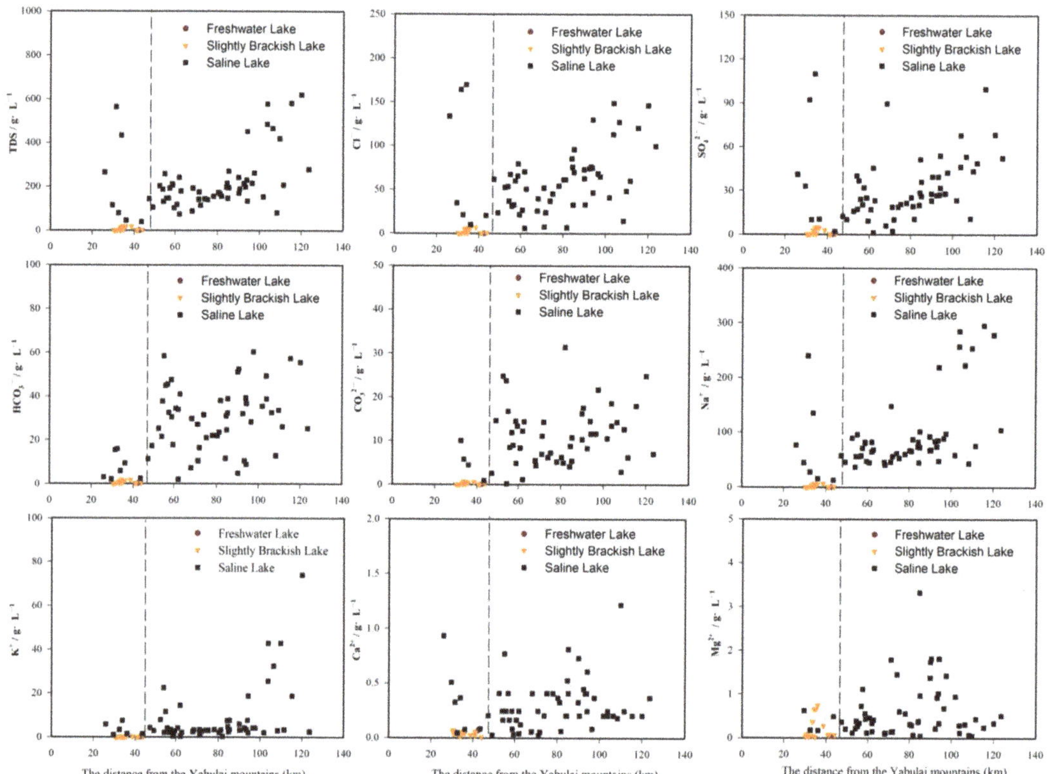

Figure 8. The variability of lake ions distance from the Yabulai Mountains.

4.3. Analysis of Mechanisms Contributing to the Hydrochemistry of Different Types of Lakes

In general, the major sources of dissolved ions in lakes include input from precipitation, weathering rocks, regional geologic, and anthropogenic input [48]. Evidenced by Gibbs diagram in Figure 4, indicated that the evaporation-crystallization reactions are the dominant in the study region. Moreover, some saline mineral deposits, including gypsum, hailte, anhydrite, epsomite, mirabilite, and thenardite, are extensive in these regions caused by intense evaporation-crystallization. Intense evaporation is a vital driving force in the formation of lakes and also affects their hydrochemical components and water balance. The higher HCO_3^-, SO_4^{2-}, Ca^{2+}, and Mg^{2+} concentration in lakes from the freshwater-lake demonstrate that those regions are enriched in minerals like anhydrite, aragonite, calcite, dolomite and gupsum. In comparison to this lake, the slightly brackish and saline lakes have lower Ca^{2+}. Calcium consumption led to the progressive enrichment of Na^+, SO_4^{2-}, and Cl^- in the lake brines where halite, thenardite, epsomiteand, and Mirabilite precipitated. The samples from lake were far below saturation with chloride minerals such as halite, showed by the negative SI values ranging between −11.36 and 3.22, which confirms that the soluble component Na^+ and Cl^- concentration was not limited by mineral equilibrium in the sediments of the study area. Table 2 showed the saturation index of main minerals in lakes water, which were used to understand the hydrogeochemical evolution processes during flow paths of the groundwater from the recharge zone to the discharge zone. Based on mineral SI result, Ca^{2+}/Na^+ ratio and the geological condition of the study area, the hydrogeochemical processes mainly included the dissolution of anhydrite, gypsum, halite, epsomite, mirabilite, and thenardite, and the precipitation of calcite, dolomite, magnesite, and aragonite from the recharge zone to the discharge zone.

5. Conclusions

The Badan Jaran desert is a vast arid region endowed with perennial lakes with varying salinities (0.88–619.74 g L^{-1}) and high pH values (7.54–10.52) due to the abundance of Na^+, K^+, Cl^- and SO_4^{2-}. The results according to TDS of the water samples from 80 lakes indicated that freshwater, slightly brackish, brackish, and saline water lakes accounted for 1.25%, 25%, 0.0%, and 73.75%, respectively. The Piper termary diagram indicate that the lake hydrochemistry changed from the SO_4^{2-}-Cl^--HCO_3^- type to the SO_4^{2-}-Cl^- type and lakes transitioned from freshwater to saline. Enrichment of Na^+ in the lake water of the Badain Jaran Desert exceeds that of Cl^-, suggesting strong evaporative concentration and cation exchange. The spatial distribution of ion contents in the lakes indicated that freshwater and slightly brackish lakes were mainly distributed in the piedmont area at a high altitude (1200–1300 m) and were concentrated within 30–45 km from the Yabulai Mountain, whereas saline lakes were mainly distributed in the desert hinterland at a low altitude (1100–1200 m) and were widely distributed over a range of between 20–135 km, there is a roughly increasing trend of ions from the Yabulai Mountains. Evidenced by Gibbs diagram indicated that the evaporation-crystallization reactions are the dominant in the study region. Moreover, some saline mineral deposits, including gypsum, hailte, anhydrite, epsomite, mirabilite, and thenardite, are extensive in these regions caused by intense evaporation-crystallization.

Author Contributions: Conceptualization, B.J. and J.S.; methodology, B.J., H.X.; formal analysis, B.J.; investigation, B.J. and J.Q.; writing-original draft preparation, B.J.; writing-review and editing, B.J., J.S.; funding acquisition, J.S. All authors have read and agreed to the published version of the manuscript.

Funding: This research was funded by the Major Science and Technology Project in Inner Mongolia Autonomous Region of China (No. zdzx2018057) and the Innovation Cross Team Project of Chinese Academy of Sciences, CAS (No. JCTD-2019-19), Transformation Projects of Scientific and Techological Achievements in Inner Mongolia Autonomous region of China (No. 2021CG0046).

Institutional Review Board Statement: Not applicable.

Informed Consent Statement: Not applicable.

Data Availability Statement: The data that support the findings of this study are available from the corresponding author on reasonal request.

Acknowledgments: We thank all the participants in the Alashan Desert Eco-hydrological Experimental research Station, Northwest Institute of Eco-Environment and Resources, Chinese Academu of Sciences. We greatly appreciate suggestions from anonymous referees for the improvement of our paper. Thanks also to the editorial staff.

Conflicts of Interest: The authors declare no conflict of interest.

References

1. Fan, Z.; Zhang, L. Hydrochemistry of lakes in Xinjiang. *Arid Zone Res.* **1992**, *9*, 1–8.
2. Ma, N.; Wang, N.; Li, Z.; Chen, X.; Zhu, J.; Dong, C. Analysis on Climate Change in the Northern and Southern Marginal Zones of the Badain Jaran Desert during the Period 1960–2009. *Arid Zone Res.* **2011**, *28*, 243–250.
3. Ma, N.; Wang, N.; Zhu, J.; Chen, X.; Chen, H.; Dong, C. Climate Change around the Badain Jaran Desert in Recent 50 Years. *J. Desert Res.* **2011**, *31*, 1541–1547.
4. Gulbostan, T.; Pang, Z.; Shang, Y. The Source of Lake Water Supply from the Comparison of Isotopic and Hydrochemical in Desert Area: Taking Mukainao Lake in Ordos Basin as an Example. *Xinjiang Geol.* **2020**, *38*, 546–551.
5. Wang, S.; Dou, H.; Chen, K.; Wang, X.; Jiang, J. *A General Survey of Lakes in China. Lakes in China*, 1st ed.; Science Press: Beijing China, 1998; Volume 6, pp. 320–347.
6. Lu, Y.; Wang, N.; Li, G.; Li, Z.; Dong, C.; Lu, J. Spatial distribution of lakes in Badain Jaran Desert. *J. Lake Sci.* **2010**, *22*, 774–782.
7. Li, J.; Edmunds, W.M.; Lu, Z.; Ma, J. Geochemistry of sediment moisture in the Badain Jaran desert: Implications of recent environmental changes and water-rock interaction. *Appl. Geochem.* **2015**, *63*, 235–247.
8. Chen, J.S.; Li, L.; Wang, J.Y.; Barry, D.; Sheng, X.F.; Zu Gu, W.; Zhao, X.; Chen, L. Groundwater maintains dune landscape. *Nature* **2004**, *432*, 459–460. [CrossRef]

9. Chen, J.; Fan, Z.; Wang, J.; Gu, W.; Zhao, X. Isotope Methods for Studying the Replenishment of the Lakes and Downstream Groundwater in the Badain Jaran Desert. *Acta Geosci. Sin.* **2003**, *24*, 497–504.
10. Ma, J.; Chen, F.; Zhao, H. Vadose geochemical records of groundwater recharge and climate change in the Badain Jaran Desert since 1000 years. *Chin. Sci. Bull.* **2004**, *49*, 22–26.
11. Ma, J.; Huang, T.; Ding, Z.; Edmunds, W. Environmental Isotopes as the Indicators of the Groundwater Recharge in the South Badain Jaran Desert. *Adv. Earth Sci.* **2007**, *22*, 922–930.
12. Chen, W.; Huang, X.; Zhou, W. *Methods of Lake Ecosystem Observation*; China Environmental Science Press: Beijing, China, 2005; ISBN 978780163745.
13. Dong, Z.; Qian, G.; Lv, P.; Hu, G. Investigation of the sand sea with the tallest dunes on Earth: China's Badain Jaran Sand Sea. *Earth-Sci. Rev.* **2013**, *120*, 20–39. [CrossRef]
14. Ma, R.; Yang, G.; Duan, H.; Jiang, J.; Wang, S.; Feng, X.; Li, A.; Kong, F.; Xue, B.; Wu, J.; et al. The number, area and spatial distribu-tion of lakes in China. *Sci. Sin. Terrae* **2011**, *41*, 394–410.
15. Zhang, Z.; Wang, N.; Ma, N.; Dong, C.; Chen, L.; Shen, S. Lakes Area change in Badain Jaran Desert Hinterland and Its Influence Factors during the Recent 40 Years. *J. Desert Res.* **2012**, *32*, 1743–1750.
16. Banda, J.F.; Lu, Y.; Hao, C.; Pei, L.; Du, Z.; Zhang, Y.; Wei, P.; Dong, H. The Effects of Salinity and pH on Microbial Commu-nity Diversity and Distribution Pattern in the Brines of Soda Lakes in Badain Jaran Desert, China. *Geomicrobiol. J.* **2020**, *37*. [CrossRef]
17. Dong, G.; Gao, Q.; Zou, X.; Li, B.; Yan, C. Climate change in southern Margin of Badain Jaran Desert since late Pleistocene. *Chin. Sci. Bull.* **1995**, *40*, 1214–1218.
18. Cao, L.; Nie, Z.; Jiang, G.; Liu, M.; He, P.; Tong, L.; Wang, J. Water Volume Change of Lakes in the Badain Jaran Desert and Its Causes Based on GF Satellite Remote Sensing Date. *Yellow River* **2020**, *42*, 40–45.
19. Yang, X.; Ma, N.; Dong, J.; Zhu, B.; Xu, B.; Ma, Z.; Liu, J. Recharge to the Inter-Dune Lakes and Holocene Climatic Changes in the Badain Jaran Desert, Western China. *Quat. Res.* **2010**, *73*, 10–19. [CrossRef]
20. Zhang, Z.; Liang, A.; Zhang, C.; Dong, Z. Gobi deposits play a significant role as sand sources for dunes in the Badain Jaran Desert, Northwest China. *CATENA* **2021**, *206*, 105530. [CrossRef]
21. Yang, X.; Williams, M.A. The ion chemistry of lakes and late Holocene desiccation in the Badain Jaran Desert, Inner Mongolia, China. *CATENA* **2003**, *51*, 45–60. [CrossRef]
22. Yang, X. Water chemistry of the lakes in the Badain Jaran Desert and their holocene evolutions. *Quat. Sci.* **2002**, *2*, 97–104.
23. Ning, K.; Wang, N.; Yang, Z.; Zhang, L.; Wang, Y.; Li, Z.; Bi, Z. Holocene vegetation history and environmental changes inferred from pollen records of a groundwater recharge lake, Badain Jaran Desert, northwestern China. *PalaGeogr. Palae-Oclimatol. Palaeoecol.* **2021**, *577*, 110538. [CrossRef]
24. Gates, J.B.; Edmunds, W.M.; Darling, W.G.; Ma, J.; Pang, Z.; Young, A.A. Conceptual model of recharge to southeastern Badain Jaran Desert groundwater and lakes from environmental tracers. *Appl. Geochem.* **2008**, *23*, 3519–3534. [CrossRef]
25. Wu, Y.; Wang, N.; Zhao, L.; Zhang, Z.; Chen, L.; Lu, Y.; Lü, X.; Chang, J. Hydrochemical characteristics and recharge sources of Lake Nuoertu in the Badain Jaran Desert. *Chin. Sci. Bull.* **2014**, *59*, 886–895. [CrossRef]
26. Zhang, J.; Wang, X.; Hu, X.; Lu, H.; Ma, Z. Research on the recharge of the lakes in the Badain Jaran Desert: Simulation study in the Sumu Jaran Lakes area. *J. Lake Sci.* **2017**, *29*, 467–479.
27. Thomas, J.; Joseph, S.; Thrivikramji, K. Hydrochemical variations of a tropical mountain river system in a rain shadow region of the southern Western Ghats, Kerala, India. *Appl. Geochem.* **2015**, *63*, 456–471. [CrossRef]
28. Han, G.; Liu, C. Hydrogeochemistry of Wujiang River water in Guizhou Province, China. *Chin. J. Geochem.* **2001**, *20*, 240–248. [CrossRef]
29. Wu, E. Hydrochemistry of inland rivers in the north Tibetan Plateau: Constrains and weathering rate estimation. *Sci. Total Environ.* **2016**, *541*, 461–482. [CrossRef]
30. Cui, B.-L.; Li, X.-Y.; Wei, X.-H. Isotope and hydrochemistry reveal evolutionary processes of lake water in Qinghai Lake. *J. Great Lakes Res.* **2016**, *42*, 580–587. [CrossRef]
31. Wang, F.; Sun, D.; Chen, F.; Bloemendal, J.; Guo, F.; Li, Z.; Zhang, Y.; Li, B.; Wang, X. Formation and evolution of the Badain Jaran Desert, North China, as revealed by a drill core from the desert center and by geological survey. *PalaGeogr. Palae-Oclimatol. Palaeoecol.* **2015**, *426*, 139–158. [CrossRef]
32. Wang, T. Formation and Evolution in the Badan Jaran Desert. *J. Desert Res.* **1990**, *10*, 29–40.
33. Wang, X.; Hu, X.; Jin, X.; Hou, L.; Qian, R.; Wang, L. Interactions between groundwater and lakes in the Badan Jaran Desert. *Earth Sci. Front.* **2014**, *21*, 91–99.
34. Zhang, J.; Wang, X.; Hu, X.; Lu, H.; Gong, Y.; Wan, L. The macro- characteristics of groundwater flow in the Badain Jaran desert. *J. Desert Res.* **2015**, *35*, 774–782.
35. Rui, X. *The Surface Water. Principles of Hydrology*, 1st ed.; China Water & Power Press: Beijing, China, 2004; ISBN 9787508421643.
36. Shao, T.; Zhao, J.; Dong, Z. Water Chemistry of the Lakes and Groundwater in the Badain Jaran Desert. *Acta Geogr. Sin.* **2011**, *66*, 662–670.
37. Wang, X.; Zhou, Y. Investigating the mysteries of grounds in the Badain Jaran Desert, China. *Hydrogeol. J.* **2018**, *26*, 1639–1655. [CrossRef]
38. Dong, C.; Wang, N.; Chen, J.; Li, Z.; Chen, H.; Chen, L.; Ma, N. New obervational and experimental evidence for the re-charge mechanism of the lake group in the Alxa Desert, North-central China. *J. Arid Environ.* **2016**, *124*, 48–61. [CrossRef]

39. Gao, Z.; Lin, Z.; Niu, F.; Luo, J.; Liu, M.; Yin, G. Hydrochemistry and controlling mechanism of lakes in permafrost regions along the Qinghai-Tibet Engineering Corridor, China. *Geomorphology* **2017**, *297*, 159–169. [CrossRef]
40. He, J.; Zhang, J.; Ding, Z.; Ma, J. Evolution of groundwater geochemistry in the Minqin Basin: Taking chloride as an indica-tor of groundwater recharge. *Resour. Sci.* **2011**, *33*, 416–421.
41. He, J.; Ma, J.; Zhang, P.; Tian, L.; Zhu, G.; Edmunds, W.M.; Zhang, Q. Groundwater recharge environments and hydrogeo-chemical evolution in the Jiuquan Basin, Northwest China. *Appl. Geochem.* **2012**, *27*, 866–878. [CrossRef]
42. Ma, N.; Yang, X. Characteristics of hydrochemistry and environmental isotopes in the Badain Jaran Desert and its southeast-ern margin and their hydrological significance. *Quat. Sci.* **2008**, *28*, 702–711.
43. Zhang, H.; Hu, X.; Wang, X. Research on Geomorphologic Evolution of East and West Badain Lake and the Impact on Water Body Features. *Geoscience* **2017**, *31*, 406–414.
44. Jin, K.; Rao, W.; Zhang, W.; Zheng, F.; Wang, S. Strontium isotopic and hydrochemical characteristics of shallow groundwater and lake water in the Badain Jaran Desert, North China. *Isot. Environ. Health Stud.* **2021**, *57*, 516–534. [CrossRef]
45. Ma, J.; Enmunds, W. Groundwater and lake evolution in the Badain Jaran Desert ecosystem, Inner Mongolia. *Hydrogeol. J.* **2006**, *14*, 1231–1243. [CrossRef]
46. Yang, X. Chemistry and late Quaternary evolution of ground and surface waters in the area of Yabulai Mountains, western Inner Mongolia, China. *CATENA* **2006**, *66*, 135–144. [CrossRef]
47. Hofmann, J. The lakes in the SE part of Badain Jaran Shamo, their limnology and geochemistry. *Geowissenschaften* **1996**, *14*, 275–278.
48. Yuan, R.; Wang, M.; Wang, S.; Song, X. Water transfer imposes hydrochemical impacts on groundwater by altering the in-teraction of groundwater and surface water. *J. Hydrol.* **2020**, *583*. [CrossRef]

Article

Topsoil Nutrients Drive Leaf Carbon and Nitrogen Concentrations of a Desert Phreatophyte in Habitats with Different Shallow Groundwater Depths

Bo Zhang [1,2,3,*], Gangliang Tang [1,3,4], Hanlin Luo [1,3,5], Hui Yin [1,3,4], Zhihao Zhang [1,3,4], Jie Xue [1,3,4], Caibian Huang [1,3,4], Yan Lu [1,3,4], Muhammad Shareef [1,3,4], Xiaopeng Gao [6] and Fanjiang Zeng [1,3,4,*]

1 Xinjiang Key Laboratory of Desert Plant Roots Ecology and Vegetation Restoration, Xinjiang Institute of Ecology and Geography, Chinese Academy of Sciences, Urumqi 830011, China; tanggangliang@ms.xjb.ac.cn (G.T.); huangmozhixing@163.com (H.L.); yinhui@ms.xjb.ac.cn (H.Y.); zhangzh@ms.xjb.ac.cn (Z.Z.); xuejie11@ms.xjb.ac.cn (J.X.); huangcaibian@ms.xjb.ac.cn (C.H.); luyan@ms.xjb.ac.cn (Y.L.); shareefagron@gmail.com (M.S.)
2 National Engineering Technology Research Center for Desert-Oasis Ecological Construction, Xinjiang Institute of Ecology and Geography, Chinese Academy of Sciences, Urumqi 830011, China
3 Cele National Station of Observation and Research for Desert-Grassland Ecosystems, Cele 848300, China
4 State Key Laboratory of Desert and Oasis Ecology, Xinjiang Institute of Ecology and Geography, Chinese Academy of Sciences, Urumqi 830011, China
5 Lanzhou New District Urban Development Investment Group Co., Ltd., Lanzhou 730000, China
6 Department of Soil Science, University of Manitoba, Winnipeg, MB R3T 2N2, Canada; gaoxp@umanitoba.ca
* Correspondence: zhangbo@ms.xjb.ac.cn (B.Z.); zengfj@ms.xjb.ac.cn (F.Z.)

Citation: Zhang, B.; Tang, G.; Luo, H.; Yin, H.; Zhang, Z.; Xue, J.; Huang, C.; Lu, Y.; Shareef, M.; Gao, X.; et al. Topsoil Nutrients Drive Leaf Carbon and Nitrogen Concentrations of a Desert Phreatophyte in Habitats with Different Shallow Groundwater Depths. *Water* **2021**, *13*, 3093. https://doi.org/10.3390/w13213093

Academic Editor: Maria Mimikou

Received: 23 September 2021
Accepted: 31 October 2021
Published: 3 November 2021

Abstract: Phreatophytes are deep-rooted plants that reach groundwater and are widely distributed in arid and semiarid areas around the world. Multiple environmental factors affect the growth of phreatophytes in desert ecosystems. However, the key factor determining the leaf nutrients of phreatophytes in arid regions remains elusive. This study aimed to reveal the key factors affecting the ecological stoichiometry of desert phreatophytes in the shallow groundwater of three oases at the southern rim of the Taklimakan Desert in Central Asia. Groundwater depth; groundwater pH and the degree of mineralization of groundwater; topsoil pH and salt concentration; topsoil and leaf carbon, nitrogen, and phosphorus concentrations of phreatophytic *Alhagi sparsifolia* grown at groundwater depths of 1.3–2.2 m in the saturated aquifer zone in a desert–oasis ecotone in northwestern China were investigated. Groundwater depth was closely related to the mineralization degree of groundwater, topsoil C and P concentrations, and topsoil salt content and pH. The ecological stoichiometry of *A. sparsifolia* was influenced by depth, pH and the degree of mineralization of groundwater, soil nutrients and salt concentration. However, the effects of soil C and P concentrations on the leaf C and N concentrations of *A. sparsifolia* were higher than those of groundwater depth and pH and soil salt concentration. Moreover, *A. sparsifolia* absorbed more N in the soil than in the groundwater and atmosphere. This quantitative study provides new insights into the nutrient utilization of a desert phreatophyte grown at shallow groundwater depths in extremely arid desert ecosystems.

Keywords: ecological stoichiometry; *Alhagi sparsifolia*; groundwater table; soil salt; extremely arid region; ecological protection

1. Introduction

Carbon, nitrogen, and phosphorus are vital microelements in most terrestrial plants. C, which comes from CO_2 in the air by plant photosynthesis, is the foundation for synthesizing nearly all organic compounds. Nitrogen is a critical component of proteins, nucleic acids, nucleotides, and enzymes as well as amino acid synthesis. Phosphorus is a vital element of phospholipids, water-soluble P esters, phosphoproteins, DNA, and RNA [1,2]. C, N, and P concentrations in plants are strongly related to plant growth status [1], and the N:P ratio in plants is widely used to indicate nutrient limitation in various ecosystems [3,4]. In addition, plant elemental content and ratio variations are closely related to ecosystem processes such as productivity, energy flow, plant growth strategy, and decomposition [5,6]. Therefore, plant elemental content and their ratio variations are useful tools for understanding the mechanisms of plant growth status and ecosystem functions in different ecosystems.

Plant roots that access groundwater to obtain water can be considered phreatophytes. Phreatophytes are a hydroecological plant type that is widely distributed in arid and semiarid areas in grasslands and deserts around the world, except Antarctica [7]. Through their deep roots, phreatophytes have remarkable advantages in regions with insufficient precipitation. Therefore, phreatophytes can adapt better to arid environments than other plants [5,8,9]. *Alhagi sparsifolia* Shap., a dominant perennial phreatophyte grown at the southern rim of the Taklimakan Desert, has an important influence on wind prevention and sand fixation in the local ecosystem. *A. sparsifolia* is a leguminous plant and is an economically important species for local livestock because of its high protein concentration [5,8,10,11]. The deep roots of *A. sparsifolia* usually reach 20 m to meet groundwater, which is the main source of water and nutrients [12,13].

Numerous studies have identified several factors affecting plant growth and nutrient concentrations in *A. sparsifolia* [5,8,10,14–21]. First, previous studies have shown that groundwater is a major source of nutrients and water for *A. sparsifolia*. The xylem sap concentrations of NO_3 and PO_3 in *A. sparsifolia* indicate a sufficient nutrient supply via groundwater [7,12,13]. Therefore, the variation in groundwater depth has a significant effect on the biomass, cover, growth rate, leaf P concentration, and nutrient resorption of *A. sparsifolia* [5,9,15,16]. Second, high groundwater salinity remarkable alters biomass and species diversity [19,22,23]. The biomass of *A. sparsifolia* declines with increasing groundwater salinity on shallow groundwater [23], and low groundwater salinity is closely related to high plant diversity [19]. Third, soil nutrients influence plant growth and leaf nutrients of *A. sparsifolia* in different natural habitats [5,8,10,17]. The leaf N and P concentrations of *A. sparsifolia* are positively related to soil N and P concentrations [5,9,10,17]. Soil nutrients are the main driving factors affecting the bacteria and fungus communities of *A. sparsifolia* in arid desert ecosystems [20]. Finally, soil salinity notably affects the population density and nitrate assimilation of *A. sparsifolia*. The population density of *A. sparsifolia* is approximately 20% at a slight soil salinity and drops to less than 15% under severe soil salinity [14]. The mineral nitrogen assimilation of *A. sparsifolia* is more sensitive to soil salinity than biological N fixation [21]. *A. sparsifolia* rehabilitates saline soil environments because of its high ion selectivity, which excludes Na^+ and accumulates Ca^{2+} in its leaves [18]. However, the key factor that affects leaf C, N, and P concentrations in *A. sparsifolia* is still unknown.

The Taklimakan Desert, which covers 37,000 square kilometers, is the world's second-largest shifting sand desert and China's largest desert. Previous reports have shown that phreatophytic desert plants always reach the groundwater to obtain water and nutrient resources and make them dependent on water and nutrients from soil in the Taklimakan Desert [7,12,13,21]; however, studies revealing the most important factor that determines the leaf nutrients of phreatophytes are relatively few. In this study, we chose five study sites in three desert–oasis ecotones at the southern rim of the Taklimakan Desert to determine the environmental factors, such as groundwater depth and salinity, soil nutrients, and salinity, that play a crucial role in influencing the leaf C, N, and P concentrations of *A. sparsifolia* in an arid ecosystem. Our hypothesis was that groundwater depth has the most important effect on the variation of nutrients in *A. sparsifolia* in various shallow groundwater habitats.

2. Materials and Methods

2.1. Study Area

The study sites were located in Cele, Moyu, and Hetian counties in the Hetian Prefecture (79°28′ E–81°04′ E, 36°54′ N–37°14′ N), which belong to the southern rim of the Taklimakan Desert of the Xinjiang Uyghur Autonomous Region, China (Figure S1). The Taklimakan Desert is located in the Tarim Basin in western China and in the central part of

Asia within the rain shadows of the surrounding Tian Shan and Kunlun mountain ranges. Consequently, the climate is extremely arid in this area. The mean annual precipitation ranges from 20 to 70 mm, and the annual potential evapotranspiration is approximately 2600 mm. The highest and the lowest temperatures are 41.9 °C in July and −23.9 °C in January, respectively [9]. Vegetation, including the herbaceous legume *Alhagi sparsifolia* Shap. (Fabaceae), herbaceous perennial *Karelinia caspia* (Pall.) Less. (Asteraceae), salt cedar *Tamarix ramosissima* Ledeb. (Tamaricaceae), *Phragmites australis* (Cav.) *Trin. ex Steu.* (Poaceae), and poplar tree *Populus euphratica* Oliv. (Salicaceae), is sparse in the five research sites. The groundwater depth ranges from 1.3 to 2.2 m in the saturated aquifer zone. The soils are aeolian sandy soil in these five research sites. The soil is Aridisol in the USDA ST system and contains 90% sand, 4% clay, and 6% silt at a 0–20 cm soil depth [24].

2.2. Methods

Each site had four quadrats that covered 10 × 10 m. During the experimental period from May to August 2016, four samples of soil (20 cm depth), groundwater, and plant samples were collected monthly. The samples were dried to a constant weight and maintained at 70 °C. After natural air drying, soil samples were treated using the quartering method and a 2 mm soil sieve. Soil water content was determined using a drying method. Soil salinity was measured using the residue-drying method. Air-dried soil samples were passed through a 2 mm sieve, and a 10.0 g soil sample was added to 50 mL ultrapure water (soil: water ratio = 1:5). The mixture was shaken for 3 min. Then, 20 mL of the supernatant was placed in a weighed beaker and dried [25]. Soil pH was assayed using the Mettler Toledo S20K pH meter (Mettler Toledo Instruments, Greisensee, Switzerland). Soil organic carbon was determined using the $K_2Cr_2O_7$ digestion method. Total N and P contents were determined using the Kjeldahl and digestion methods, respectively [8]. Groundwater pH was assayed using a Mettler Toledo S20K pH meter (Mettler Toledo Instruments, Greisensee, Switzerland). The degree of mineralization of the groundwater was determined using a drying method. Groundwater (100 mL) was placed in a small beaker and dried in a steam bath. If a colored steaming residue was observed, the sample was added with a few drops of (1 + 1) hydrogen peroxide solution, dried in a steam bath until the color of the residue did not change, baked at 105 °C for 2 h, and weighed [26]. The leaf organic content was determined using a total organic carbon analyzer (Aurora 1030; College Station, TX, USA). Leaf N and P contents were determined using the Kjeldahl and digestion methods, respectively [8].

2.3. Data Analysis

One-way ANOVA with Tukey's test was carried out for the multiple comparisons of groundwater depth, groundwater pH, and the degree of mineralization of groundwater; soil salt, soil pH, and C, N, and P concentrations of soil and *A. sparsifolia* at different research sites to test statistical significance ($p < 0.05$). Linear relationships among variables (i.e., soil, plant, and groundwater) were analyzed using the "corrplot" package in R software (version 4.0.3). The statistical significance values of each explanatory variable, including groundwater and soil variables, were tested using the "rdacca.hp" package in the R software [27]. The structural equation model (SEM) was used to tease apart statistically and quantify the direct and indirect effects of groundwater and soil variables on the leaf nutrient concentration using the R software with the package "piecewiseSEM" [9,28]. The D-separation test of piecewise SEM was used to test whether the causal model missed important links, and a $p > 0.05$ indicated that the model was acceptable [29].

3. Results

3.1. Groundwater and Soil Characteristics

The mean and standard deviation for depth, pH, and the degree of mineralization of groundwater as well as soil pH, salt, C, N, and P for the different sites are shown in Figure 1. Significant variations in groundwater and soil characteristics were observed at different study sites (Figure 1). The depth and pH of groundwater ranged from 1.30 to 2.20 m (Figure 1A) and 7.03 to 8.84 (Figure 1B), respectively. Groundwater depth was the highest in Hetian county (HT) and the lowest in northern Cele County (CN) among all study sites. The degree of mineralization of groundwater was the highest in CN and the lowest in the west of Cele County (CW) and HT among the five study sites. No significant differences were observed in the pH of the groundwater in this study. In addition, the soil pH was highest in HT and lowest in CW, Moyu County (MY), and CN among the five study sites. The soil salt concentration in CN was higher than that in the other study sites. Soil C concentrations were the highest in CN and the east of Cele County (CE) and the lowest in CW and HT, soil N concentrations were the highest in CN and the lowest in HT, and soil P concentrations were the highest in HT and the lowest in CW.

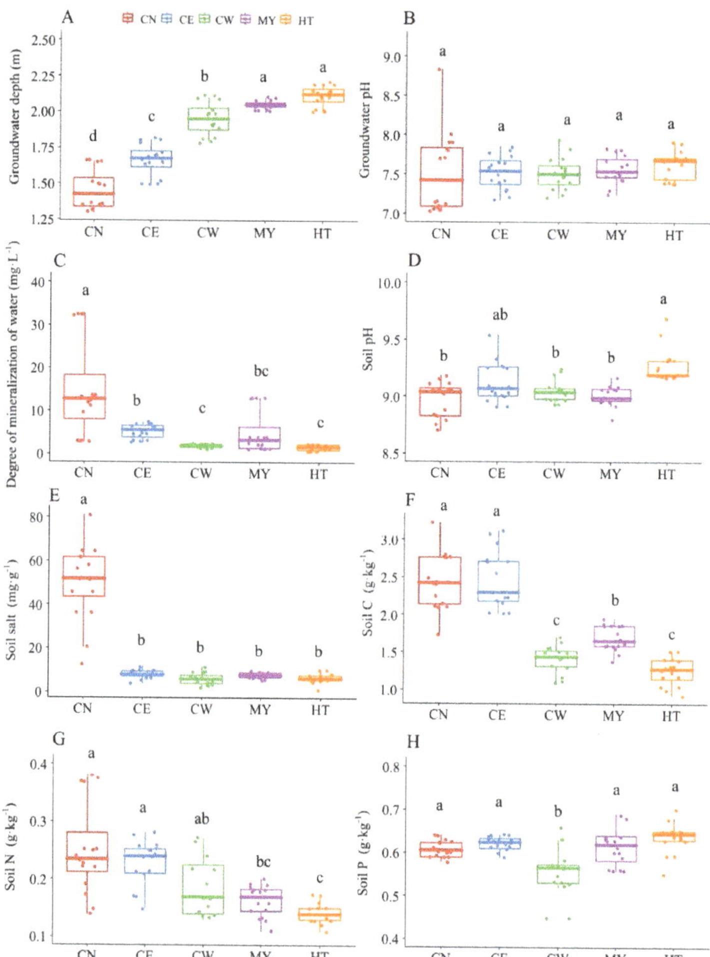

Figure 1. Changes in the environmental variables of the study sites. Vertical bars represent standard deviations, and the letters denote significant differences among different groundwater depths ($p < 0.05$). CW, west of Cele county; MY, Moyu county; HT, Hetian county; CN, north of Cele county; CE, east of Cele county. (**A–H**) represent variations of depth, pH, and the mineralization degree of groundwater as well as pH, salt, C, N, and P concentrations of soil.

3.2. Variations in Leaf Ecological Stoichiometry

The mean and standard deviation for leaf C, N, P, and the N:P ratio of *A. sparsifolia* for the different sites are shown in Figure 2. Significant differences in the leaf ecological stoichiometry of *A. sparsifolia* were observed at the study sites (Figure 2). In all study sites, leaf C concentrations were the highest in CE and the lowest in CW and HT; leaf N concentrations were the highest in CW and the lowest in HT; leaf N:P ratios were the highest in CW and the lowest in HT, CN, and CE. No significant differences were observed in the leaf P concentrations.

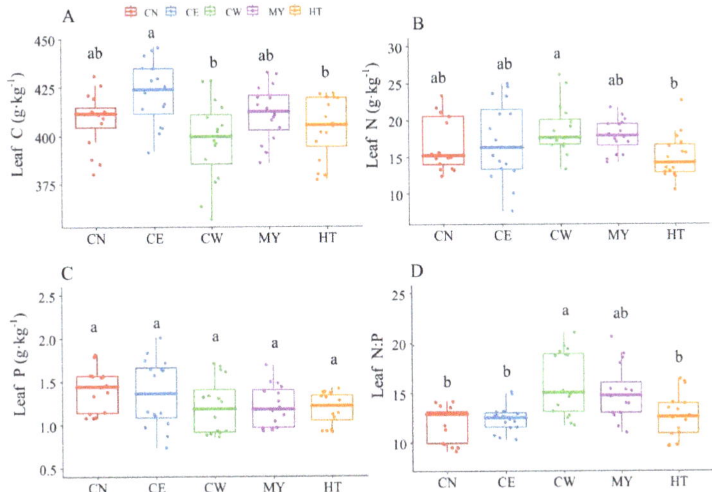

Figure 2. Changes in the leaf ecological stoichiometry of *A. sparsifolia* at the study sites. Vertical bars represent standard deviations, and the letters denote significant differences among different groundwater depths ($p < 0.05$). CW, west of Cele county; MY, Moyu county; HT, Hetian county; CN, north of Cele county; CE, east of Cele county. (**A–D**) represent variations of leaf C, N, P, and N:P ratio at the study sites.

3.3. Relationships among Groundwater, Soil, and Leaf Ecological Stoichiometry

Significant relationships ($p < 0.05$) were observed among groundwater depth, groundwater salinity, soil nutrient concentration, and soil salt concentration as well as leaf ecological stoichiometry of *A. sparsifolia* (Figure 3). Groundwater depth was significantly related with leaf P concentration and leaf N:P ratio. In addition, groundwater depth was significantly related to the degree of mineralization of groundwater, soil pH, soil salt, and soil C and N concentrations. A significant positive correlation was observed between the groundwater pH and leaf P concentration. Soil C concentration was strongly correlated to leaf P concentration and leaf N:P ratio. The soil N concentration was closely linked to leaf N:P ratio. The soil P concentration was negatively related to leaf N concentration. Soil salt concentration had a significant impact on leaf N:P ratio.

3.4. Relative Influences of Environmental Factors on Variation in Leaf Ecological Stoichiometry

The leaf C concentration of *A. sparsifolia* was largely determined by groundwater pH and soil P concentration (Figure 4A). Groundwater pH accounted for 39.9% of the variation in leaf C concentration followed by 18.9% variation in soil P concentration. The leaf P concentration was significantly influenced by groundwater table depth and soil C concentration, which explained 40.3% and 35.0% of variations in leaf P concentration, respectively (Figure 4C). The leaf ecological stoichiometry was strongly influenced by groundwater pH and soil P concentration, which contributed to 38.9% and 18.8% of the variations, respectively, of leaf ecological stoichiometry (Figure 4D). No significant environmental factors affected leaf N concentrations (Figure 4B).

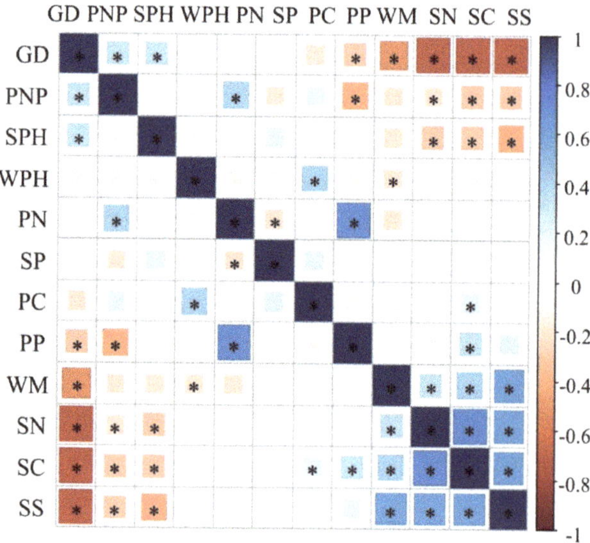

Figure 3. Relationships between groundwater, soil, and leaf ecological stoichiometry of *A. sparsifolia* at the study sites. GD, groundwater depth; WPH, groundwater pH; WM, degree of mineralization of groundwater; SS, soil salt; SPH, soil pH; SC, soil C concentration; SN, soil N concentration; SP, soil P concentration; PC, plant C concentration; PN, plant N concentration; PP, plant P concentration; PNP, plant N:P ratio. * Indicates significant correlation at $p < 0.05$.

3.5. Environmental Factors Affecting Variations in Leaf Ecological Stoichiometry

The key factors affecting the variations in leaf C, N, and P concentrations of *A. sparsifolia* were explored using SEM (Figure 5). The model explained 29%, 52%, and 15% of the variance in leaf C, N, and P concentrations, respectively. Groundwater depth had a direct effect (0.20) on leaf N concentration. Groundwater pH had direct effects (0.34) on the leaf C concentration and indirect effects (0.07) on the leaf N concentration. The soil C concentration had direct effects (0.45) on leaf C concentration and indirect effects (0.01) on leaf N concentration. In addition, the soil P concentration had a direct effect (0.21) on leaf N concentration. Therefore, the effects of soil nutrients (0.67) on variations in leaf C and N concentrations were higher than those of groundwater characteristics (0.61).

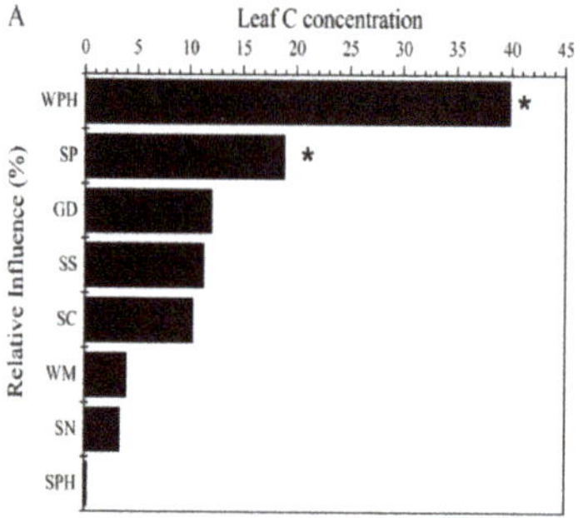

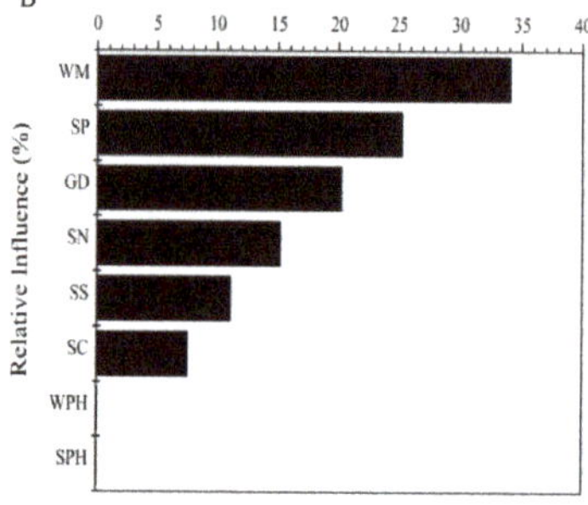

Figure 4. *Cont.*

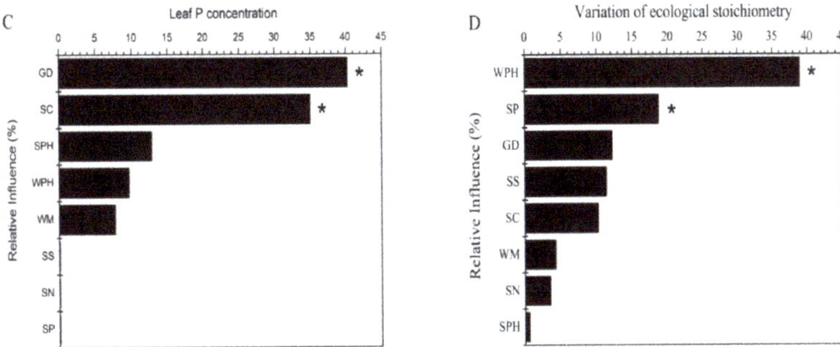

Figure 4. Relative influences of environmental factors on variations in leaf C, N, and P concentrations as well as ecological stoichiometry of *A. sparsifolia* in the study sites. GD, groundwater depth; WPH, groundwater pH; WM, the degree of mineralization of groundwater; SS, soil salt; SPH, soil pH; SC, soil C concentration; SN, soil N concentration; SP, soil P concentration; PC, plant C concentration; PN, plant N concentration; PP, plant P concentration; PNP, plant N:P ratio. * Indicates significant correlation at $p < 0.05$. (**A–D**) represent relative influences of environmental factors on variations in leaf C. N, P, and ecological stoichiometry of *A. sparsifolia*.

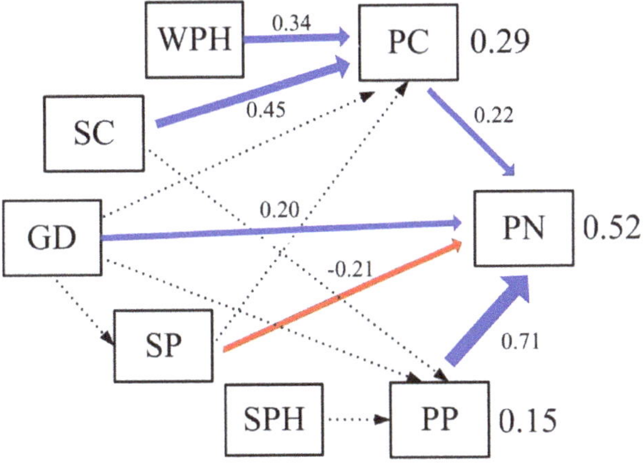

Figure 5. Controlling factors analysis of variations in leaf C, N, and P concentrations of *A. sparsifolia* in the study sites using the structural equation model. Significant regressions and nonsignificant regressions are shown by solid ($p < 0.05$) and dashed lines, respectively. The thickness of significant paths was scaled on the basis of magnitude of the standardized regression coefficient. Blue and red arrows denote positive and negative relationships, respectively. GD, groundwater depth; WPH, groundwater pH; WM, degree of mineralization of groundwater; SS, soil salt; SPH, soil pH; SC, soil C concentration; SN, soil N concentration; SP, soil P concentration; PC, plant C concentration; PN, plant N concentration; PP, plant P concentration; PNP, plant N:P ratio.

4. Discussion

4.1. Topsoil Nutrients Affected Variations in Leaf Ecological Stoichiometry

In contrast to our hypothesis, our results showed that soil P concentration had a direct negative effect on leaf N concentration in *A. sparsifolia* (Figures 3 and 5). In addition, the soil C concentration had a direct positive effect on the leaf C concentration of *A. sparsifolia* at a 1.3–2.2 m groundwater depth in the saturated aquifer zone (Figures 3 and 5). These results agreed with previous experiments that showed that the soil P concentration was the vital factor influencing the growth and plant N and P concentrations of *A. sparsifolia* at different groundwater depths in a large area [5,8]. However, our results were not in agreement with the findings of Zhang et al. [5], who reported that soil P concentration was positively correlated to the leaf N concentration of *A. sparsifolia* and that the soil C concentration was not associated with the leaf C concentration of *A. sparsifolia* at a 2.5–11.0 m groundwater depth. One possible explanation for the different relationships was that topsoil nutrients had a greater effect on the leaf C and N concentrations of *A. sparsifolia* than groundwater in shallow groundwater depth environments. A previous study showed that soil mineralizes N leached into groundwater, which is the key source of nutrients for *A. sparsifolia* [13]. In the present study, *A. sparsifolia* was sampled at a 1.3–2.2 m groundwater depth in the saturated aquifer zone, which was shallower than that studied by Zhang et al. [5], i.e., a 2.5–11.0 m groundwater depth. In shallow groundwater environments, *A. sparsifolia* might absorb mineralized N in topsoil but not in groundwater due to the different nutrient uptake strategies in the saturated aquifer zone. This finding was consistent with the observation that plants uptake water and nutrients via the root system. The horizontal roots of *A. sparsifolia* quickly expand and tiller in shallow groundwater tables [30], and the vertical root depth of *A. sparsifolia* increases with increasing groundwater depth [30,31]. In addition, Gui et al. [15] reported that the soil water content in a shallow groundwater table (2.5 m) was higher than that in deep groundwater (4.5 and 11.0 m). In addition, our research team found that *A. sparsifolia* develops different efficient root architectures that absorb soil nutrients and water in response to nutrient and water availability, respectively [32]. Hence, *A. sparsifolia* has different nutrient uptake strategies to adapt to different groundwater depths and soil conditions in a desert ecosystem.

4.2. Groundwater Depth Influenced Variations in Leaf Ecological Stoichiometry

Previous studies showed that groundwater depth significantly influenced the leaf P concentration of *A. sparsifolia* but did not affect the leaf N concentration at 2.5–11.0 m [5]. Herein, we report that the leaf N concentration in *A. sparsifolia* was positively influenced by groundwater depth and that the leaf P concentration was not altered by groundwater depth in shallow groundwater habitats (1.3–2.2 m) (Figure 5). This finding was because *A. sparsifolia* absorbed more soil N in shallow groundwater environments than in deep groundwater. In a previous study, biological N_2 fixation accounts for more than 80% of the leaf N concentration in *A. sparsifolia* [12]. However, biological N_2 fixation requires more energy than nitrate uptake in natural ecosystems [33]. Hence, we believe that *A. sparsifolia* prefers to absorb N directly in soil for N fixation by the atmosphere in shallow groundwater environments. This finding was consistent with the observation that the topsoil around *A. sparsifolia* had a higher soil N concentration in the shallow groundwater table (2.5 m) than in deep groundwater (4.5 and 11.0 m) [34]. Moreover, the hydrochemical process of groundwater is significantly affected by topsoil nitrate and DOC concentrations [35]. Therefore, groundwater depth had different effects on the leaf N and P concentrations of *A. sparsifolia* in various habitats (shallow groundwater table or deep groundwater). These results also explained why topsoil nutrients were important for leaf carbon and nitrogen concentrations of *A. sparsifolia* in shallow groundwater depth environments. This adaptive mechanism could help phreatophytes to adapt to different hostile environments and explain their wide distribution in the extremely arid Taklimakan Desert.

4.3. Groundwater pH Affected Variations in Leaf Ecological Stoichiometry

In the Taklimakan Desert, surficial waters always have a high pH and are rich in sodium chloride and sulfate concentrations [36]. Thus, although plant roots have access to groundwater, the salinity of water affects plant growth [37]. In the present study, groundwater pH was positively related to the leaf C concentration of *A. sparsifolia* (Figures 3 and 5) and was significantly associated with the degree of mineralization of groundwater (Figure 3). This finding confirmed that the degree of mineralization of groundwater is a vital factor that alters the growth of *A. sparsifolia* at shallow groundwater depths (1.3–2.2 m) in the Taklimakan Desert [38]. Herein, we found that groundwater characteristics affected the leaf nutrients of *A. sparsifolia*, but these effects were lower than those of soil nutrients. Further research is needed to elucidate the different nutrient uptake mechanisms of phreatophytes at various groundwater depths in arid desert ecosystems.

5. Conclusions

Topsoil nutrients drive leaf carbon and nitrogen concentrations in *A. sparsifolia* with shallow groundwater. In the saturated aquifer zone, topsoil nutrients have more influence on leaf nutrients of *A. sparsifolia* than groundwater depth, salinity, and soil salt concentrations. Groundwater depth and pH also affected the leaf carbon and nitrogen concentrations in *A. sparsifolia*. However, topsoil nutrients play a crucial role in influencing leaf nutrients in *A. sparsifolia* in shallow groundwater habitats. Phreatophytic *A. sparsifolia* preferred soil N absorption in shallow groundwater environments to groundwater and biological nitrogen fixation in deep groundwater. To our knowledge, our results provide a new adaptive strategy for nutrient utilization of a desert phreatophyte grown at shallow groundwater depths in the saturated aquifer zone in a desert–oasis ecotone. This study may contribute to the protection and restoration of phreatophytes in hyperarid desert ecosystems.

Supplementary Materials: The following are available online at https://www.mdpi.com/article/10.3390/w13213093/s1, Figure S1: Location of the study sites. MY, Moyu county; HT, Hetian county; CW, west of Cele county; CN, north of Cele county; CE, east of Cele county.

Author Contributions: Conceptualization, B.Z., X.G. and F.Z.; methodology, B.Z., X.G. and F.Z..; software, B.Z. and Z.Z.; investigation, B.Z., H.L., H.Y., Y.L., Z.Z. and J.X.; writing—original draft preparation, B.Z. and G.T.; writing—review and editing, C.H., Y.L., M.S., X.G. and F.Z.; All authors have read and agreed to the published version of the manuscript.

Funding: This study was funded by the Western Young Scholar Program-B of the Chinese Academy of Sciences (2018-XBQNXZ-B-018), the National Natural Science Foundation of China (31500367; 42071259), the Key Program of Joint Funds of the National Natural Science Foundation of China and the Government of Xinjiang Uygur Autonomous Region of China (U1603233), the Youth Innovation Promotion Association Foundation of the Chinese Academy of Sciences (2020435), the Third Batch of Tianshan Talents Program of Xinjiang Uygur Autonomous Region (2021–2023), and the Project for Cultivating High-Level Talent of Xinjiang Institute of Ecology and Geography, Chinese Academy of Sciences (E0502101).

Institutional Review Board Statement: Not applicable.

Informed Consent Statement: Not applicable.

Data Availability Statement: All data reported here is available from the authors upon request.

Acknowledgments: We thank Yonggang Li and Jiangshan Lai for assistance in data analysis and Mingfang Hu for assistance in soil and plant elemental analysis. We are also grateful to the anonymous referees for reviewing this manuscript.

Conflicts of Interest: The authors declare no conflict of interest.

References

1. Elser, J.J.; Acharya, K.; Kyle, M.; Cotner, J.; Makino, W.; Markow, T.; Watts, T.; Hobbie, S.; Fagan, W.; Schade, J.; et al. Growth rate-stoichiometry couplings in diverse biota. *Ecol. Lett.* **2003**, *6*, 936–943. [CrossRef]
2. Raven, J.A. Interactions between Nitrogen and Phosphorus metabolism. *Annu. Plant Rev.* **2015**, *48*, 187–214.
3. Koerselman, W.; Meuleman, A.F.M. The Vegetation N:P Ratio: A New Tool to Detect the Nature of Nutrient Limitation. *J. Appl. Ecol.* **1996**, *33*, 1441–1450. [CrossRef]
4. Güsewell, S. N: P ratios in terrestrial plants: Variation and functional significance. *New Phytol.* **2004**, *164*, 243–266. [CrossRef]
5. Zhang, B.; Gao, X.; Li, L.; Lu, Y.; Shareef, M.; Huang, C.; Liu, G.; Gui, D.; Zeng, F. Groundwater Depth Affects Phosphorus But Not Carbon and Nitrogen Concentrations of a Desert Phreatophyte in Northwest China. *Front. Plant Sci.* **2018**, *9*, 338. [CrossRef] [PubMed]

6. Hu, Y.; Liu, X.; He, N.; Pan, X.; Long, S.; Li, W.; Zhang, M.; Cui, L. Global patterns in leaf stoichiometry across coastal wetlands. *Glob. Ecol. Biogeogr.* **2021**, *30*, 852–869. [CrossRef]
7. Thomas, F.M. *Ecology of Phreatophytes. Progress in Botany*; Springer: Berlin/Heidelberg, Germany, 2014.
8. Zhang, B.; Gui, D.; Gao, X.; Shareef, M.; Li, L.; Zeng, F. Controlling Soil Factor in Plant Growth and Salt Tolerance of Leguminous Plant *Alhagi sparsifolia* Shap. in Saline Deserts, Northwest China. *Contemp. Probl. Ecol.* **2018**, *11*, 111–121. [CrossRef]
9. Zhang, B.; Tang, G.L.; Yin, H.; Zhao, S.L.; Shareef, M.; Liu, B.; Gao, X.P.; Zeng, F.P. Groundwater Depths Affect Phosphorus and Potassium Resorption but Not Their Utilization in a Desert Phreatophyte in Its Hyper-Arid Environment. *Front. Plant Sci.* **2021**, *12*, 665168. [CrossRef]
10. Zhang, B. Responses of the Nutrient Utilization of a Phreatophyte *Alhagi Sparsifolia* Shap. at Different Groundwater Depths in an Arid Region. Ph.D. Thesis, University of Chinese Academy of Sciences, Beijing, China, 2018.
11. Zeng, F.J.; Liu, B. *Root Ecology of Alhagi Sparsifolia*; Beijing, Science Press: Beijing, China, 2012.
12. Arndt, S.K.; Kahmen, A.; Arampatsis, C.; Popp, M.; Adams, M. Nitrogen fixation and metabolism by groundwater-dependent perennial plants in a hyperarid desert. *Oecologia* **2004**, *141*, 385–394. [CrossRef]
13. Zeng, F.; Bleby, T.M.; Landman, P.A.; Adams, M.; Arndt, S.K. Water and Nutrient Dynamics in Surface Roots and Soils are not Modified by Short-term Flooding of Phreatophytic Plants in a Hyperarid Desert. *Plant Soil* **2006**, *279*, 129–139. [CrossRef]
14. Jin, Q.H. Population character of *Alhagi sparsifolia* and plant community succession. *Acta Phytoecol.* **1995**, *19*, 255–260.
15. Gui, D.; Zeng, F.; Liu, Z.; Zhang, B. Characteristics of the clonal propagation of *Alhagi sparsifolia* Shap. (Fabaceae) under different groundwater depths in Xinjiang, China. *Rangel. J.* **2013**, *35*, 355–362. [CrossRef]
16. Li, C.; Zeng, F.; Zhang, B.; Liu, B.; Guo, Z.; Gao, H.; Tiyip, T. Optimal root system strategies for desert phreatophytic seedlings in the search for groundwater. *J. Arid. Land* **2015**, *7*, 462–474. [CrossRef]
17. Zhou, X.B.; Tao, Y.; Zhang, Y.M. The C, N and P stoichiometry of dominant species in different land use types in a desert-oasis ecotone of the Southern Taklimakan Desert. *Acta Prataculturae Sin.* **2018**, *27*, 15–26.
18. Min, X.; Zang, Y.; Sun, W.; Ma, J. Contrasting water sources and water-use efficiency in coexisting desert plants in two saline-sodic soils in northwest China. *Plant Biol.* **2019**, *21*, 1150–1158. [CrossRef]
19. Zeng, Y.; Zhao, C.; Shi, F.; Schneider, M.; Lv, G.; Li, Y. Impact of groundwater depth and soil salinity on riparian plant diversity and distribution in an arid area of China. *Sci. Rep.* **2020**, *10*, 7272. [CrossRef]
20. Li, W.; Jiang, L.; Zhang, Y.; Teng, D.; Wang, H.; Wang, J.; Lv, G. Structure and driving factors of the soil microbial community associated with *Alhagi sparsifolia* in an arid desert. *PLoS ONE* **2021**, *16*, e0254065.
21. Li, M.; Petrie, M.D.; Tariq, A.; Zeng, F. Response of nodulation, nitrogen fixation to salt stress in a desert legume *Alhagi sparsifolia*. *Environ. Exp. Bot.* **2020**, *183*, 104348. [CrossRef]
22. Holden, J.; Grayson, R.; Berdeni, D.; Bird, S.; Chapman, P.; Edmondson, J.; Firbank, L.; Helgason, T.; Hodson, M.; Hunt, S.; et al. The role of hedgerows in soil functioning within agricultural landscapes. *Agric. Ecosyst. Environ.* **2018**, *273*, 1–12. [CrossRef]
23. Li, X.Y. Effects of Depth to a Groundwater Table on Ecophysiological Characteristics of Dominant Species at Foreland of Oases. Ph.D. Thesis, Institute of Applied Ecology, Chinese Academy of Sciences, Beijing, China, 2004.
24. Li, Y.; Gao, X.; Tenuta, M.; Gui, D.; Li, X.; Zeng, F. Linking soil profile N_2O concentration with surface flux in a cotton field under drip fertigation. *Environ. Pollut.* **2021**, *285*, 117458. [CrossRef]
25. Fu, Z.; Wang, P.; Sun, J.; Lu, Z.; Yang, H.; Liu, J.; Xia, J.; Li, T. Composition, seasonal variation, and salinization characteristics of soil salinity in the Chenier Island of the Yellow River Delta. *Glob. Ecol. Conserv.* **2020**, *24*, e01318. [CrossRef]
26. Zhou, G.Q.; Wang, Q. Approach to determination of degree of mineralization for shallow groundwater. *Environ. Eng.* **2003**, *21*, 65–66.
27. Lai, J.S. Rdacca.hp: Hierarchical Partitioning for Redundancy Analysis and Canonical Correspondence Analysis; R Package Version 1.0–1. 2019. Available online: https://cran.r-project.org/web/packages/dplyr/index.html (accessed on 22 September 2021).
28. Lefcheck, J.S. PiecewiseSEM: Piecewise structural equation modeling in R for ecology, evolution, and systematics. *Methods Ecol. Evol.* **2016**, *7*, 573–579. [CrossRef]
29. Shipley, B. *Cause and Correlation in Biology: A User's Guide to Path Analysis, Structural Equations and Causal Inference*; Cambridge University Press: Cambridge, UK, 2002.
30. Zeng, F.; Song, C.; Guo, H.; Liu, B.; Luo, W.; Gui, D.; Arndt, S.; Guo, D. Responses of root growth of *Alhagi sparsifolia* Shap. (Fabaceae) to different simulated groundwater depths in the southern fringe of the Taklimakan Desert, China. *J. Arid. Land* **2013**, *5*, 220–232. [CrossRef]
31. Liu, B.; Zeng, F.J.; Guo, H.F.; Zeng, J. Effects of groundwater table on growth characteristics of *Alhagi sparsifolia* shap. seedlings. *Chin. J. Ecol.* **2009**, *28*, 237–242.
32. Liu, B.; Zeng, F.J.; Arndt, S.K.; He, J.X.; Luo, W.C.; Song, C. Patterns of root architecture adaptation of a phreatophytic perennial desert plant in a hyperarid desert. *South Afr. J. Bot.* **2013**, *86*, 56–62. [CrossRef]
33. Finan, T.J.; O'Brian, M.R.; Layzell, D.B.; Vessey, J.K.; Newton, W. Nitrogen fixation: Global perspectives. In Proceedings of the 13th International Congress on Nitrogen Fixation, Hamilton, ON, Canada, 2–7 July 2001.
34. Liu, Z. Root Distribution of *Alhagi Sparsifolia* Shap. Response to Different Water Conditions. Master's Thesis, Xinjiang Institute of Ecology and Geography, Chinese Academy of Sciences, Urumqi, China, 2011.
35. Medici, G.; Baják, P.; West, L.J.; Chapman, P.J.; Banwart, S.A. DOC and nitrate fluxes from farmland; impact on a dolostone aquifer KCZ. *J. Hydrol.* **2021**, *595*, 125658. [CrossRef]

36. Gibert, E.; Gentelle, P.; Liang, K.Y. Chemical and isotopic evolution of surficial waters in the Taklamakan desert (Southern Xinjiang, China). In *Isotopes in Water Resources Management, Proceedings of the A Symposium on Isotopes in Water Resources Management, Vienna, Austria, 20–24 March 1995*; IAEA: Vienna, Austria, 1996; pp. 211–212.
37. Arndt, S.K.; Arampatsis, C.; Foetzki, A.; Li, X.; Zeng, F.; Zhang, X. Contrasting patterns of leaf solute accumulation and salt adaptation in four phreatophytic desert plants in a hyperarid desert with saline groundwater. *J. Arid. Environ.* **2004**, *59*, 259–270. [CrossRef]
38. Luo, H.L.; Zeng, F.J.; Zhang, L.; Li, M.M.; Li, S.M.; Wang, B. Ecological characteristics of *Alhagi sparsifolia* in relation to environmental factors in saline habitat. *Chin. J. Ecol.* **2017**, *36*, 1801–1807.

 water

MDPI

Article

Water Supply Increases N Acquisition and N Resorption from Old Branches in the Leafless Shrub *Calligonum caput-medusae* at the Taklimakan Desert Margin

Caibian Huang [1,2,3,*], Fanjiang Zeng [1,2,3], Bo Zhang [1,2,3], Jie Xue [1,2,3] and Shaomin Zhang [4,*]

1 Xinjiang Key Laboratory of Desert Plant Roots Ecology and Vegetation Restoration, Xinjiang Institute of Ecology and Geography, Chinese Academy of Sciences, Urumqi 830011, China; zengfj@ms.xjb.ac.cn (F.Z.); zhangbo@ms.xjb.ac.cn (B.Z.); xuejie11@ms.xjb.ac.cn (J.X.)
2 State Key Laboratory of Desert and Oasis Ecology, Xinjiang Institute of Ecology and Geography, Chinese Academy of Sciences, Urumqi 830011, China
3 Cele National Station of Observation and Research for Desert-Grassland Ecosystems, Cele 848300, China
4 Institute of Nuclear Technology and Biotechnology, Xinjiang Academy of Agricultural Sciences, Urumqi 830091, China
* Correspondence: huangcaibian@ms.xjb.ac.cn (C.H.); zhangshaomin8698@126.com (S.Z.)

Abstract: Irrigation is the main strategy deployed to improve vegetation establishment, but the effects of increasing water availability on N use strategies in desert shrub species have received little attention. Pot experiments with drought-tolerant shrub *Calligonum caput-medusae* supplied with water at five field capacities in the range of 30–85% were conducted using local soil at the southern margin of the Taklimakan Desert. We examined the changes in plant biomass, soil N status, and plant N traits, and addressed the relationships between them in four- and seven-month-old saplings and mature shrubs after 28 months. Results showed that the growth of *C. caput-medusae* was highly responsive to increased soil moisture supply, and strongly depleted the soil available inorganic N pools from 16.7 mg kg^{-1} to an average of 1.9 mg kg^{-1}, although the total soil N pool increased in all treatments. Enhancement of biomass production by increasing water supply was closely linked to increasing total plant N pool, N use efficiency (NUE), N resorption efficiency (NRE), and proficiency (NRP) in four-month saplings, but that to total plant N pool, NRE, and NRP after 28 months. The well-watered plants had lower N concentrations in senesced branches compared to their counterparts experiencing the two lowest water inputs. The mature shrubs had higher NRE and NRP than saplings and the world mean levels, suggesting a higher N conservation. Structural equation models showed that NRE was largely controlled by senesced branch N concentrations, and indirectly affected by water supply, whereas NRP was mainly determined by water supply. Our results indicated that increasing water availability increased the total N uptake and N resorption from old branches to satisfy the N requirement of *C. caput-medusae*. The findings lay important groundwork for vegetation establishment in desert ecosystems.

Keywords: *Calligonum caput-medusae*; N resorption; water addition; soil inorganic N; biomass

Citation: Huang, C.; Zeng, F.; Zhang, B.; Xue, J.; Zhang, S. Water Supply Increases N Acquisition and N Resorption from Old Branches in the Leafless Shrub *Calligonum caput-medusae* at the Taklimakan Desert Margin. *Water* **2021**, *13*, 3288. https://doi.org/10.3390/w13223288

Academic Editor: Guido D'Urso

Received: 27 September 2021
Accepted: 18 November 2021
Published: 20 November 2021

1. Introduction

Approximately 10% of drylands undergoes desertification, whereas occurring areas occupy approximately 20% of the dryland population [1]. Establishing vegetation is an important tool for controlling desertification and reducing erosion in desert ecosystems [2–4]. Irrigation is the primary intervention to improve the success of vegetation establishment in desert ecosystems with low precipitation and high evapotranspiration rates [5,6]. For example, shrub planting has been a crucial strategy in the Taklimakan desert highway shelterbelt project, which crosses the largest mobile desert in China and uses drip irrigation to support vegetation to reduce wind-blown sand that blocks the road [7]. In addition to water scarcity, nitrogen (N) limitation is another primary factor controlling plant growth in desert ecosystems [8], and they generally interact and affect plant growth. Therefore, understanding the N use strategies of artificial vegetation is crucial for the success of vegetation establishment and continuous irrigation programs in extremely arid land.

Soil moisture plays an important role in regulating N mineralization and soil N availability. Reports showed that increasing water availability increased N mineralization rate and N uptake and subsequently promoted plant growth, but reduced moisture and specific soil properties (such as high soil alkalinity), thereby limiting soil N availability [9–11]. Some studies have reported that low N deposition (<6 g $Nm^{-2}{\cdot}year^{-1}$) improved the plant productivity under drought stress, but high N deposition did not [12,13]. Moreover, increases in soil N availability after N supplementation could improve plant growth and alleviate the negative effects of drought stress in arid land [14]. However, plants became mildly N constrained under sufficient moisture in the desert [15]. The latest research has shown that the annual N deposition was 0.4 g $Nm^{-2}{\cdot}year^{-1}$ in desert ecosystems of northwest China [16]. However, whether N deposition is a potential approach to mitigate N limitation for the irrigated plants in the desert is uncertain. A number of studies have reported that nutrient resorption, which is the nutrient movement from senescing tissues back to surviving tissues [17], is important especially for plants growing in infertile soils [18–20]. Generally, two approaches can be used to assess nutrient resorption, namely, resorption efficiency and proficiency. Nutrient resorption efficiency quantifies the percent of conserved nutrients in young foliage or other live parts that are translocated from senesced tissues, and resorption proficiency measures the extent to which a nutrient is withdrawn from senescing tissues [19]. Through this nutrient resorption, plants are less dependent on soil nutrient pools to maintain or increase biomass and photosynthesis [20,21]. For example, N resorption in annual plants can provide approximately 31% of N demand [22]. In forests, 45–68% of growth may depend upon resorption [23]. Furthermore, N resorption impacts litter decomposition, nutrient cycling, and resource use efficiency [24–26], thereby affecting plant productivity and nutrient cycling processes. Therefore, clarifying the potential function of NR in artificial vegetation can evaluate their potential fitness and suitability for establishment in infertile and harsh environments such as in the margins of deserts.

The relationship between N resorption and soil fertility is complex given that negative, and positive correlations with soil nutrient availability have been reported [27,28]. The role of water availability on N resorption is also under debate because some studies found that N resorption efficiency (NRE) decreases with increasing soil water availability due to the enhancement of soil nutrient release [29,30]. However, drought can cause early onset of senescence [31], potentially increasing the importance of nutrient resorption [32]. However, nutrient resorption may be sensitive to drought limitation [33,34] due to the reduced nutrient retranslocation in the phloem and water recycling in the xylem [35]. Therefore, whether N resorption in shrubs in arid environments depends on water availability and whether growth stimulation by watering increases N resorption to satisfy increased plant N demand remains unclear.

Seedling establishment of woody species in harsh and extreme environmental conditions is the most vulnerable stage in vegetation establishment [36,37]. Several studies have reported that N limitation at the seedling stage restricted vegetative growth [38], and subsequent vigorous seedling establishment [39]. Moreover, mature trees tend to be more efficient in N-recycling than younger ones [23]. Therefore, N resorption, as an important N recycling strategy, is reported to change over time [40], and its resorption efficiency is likely to increase when plant-available N is limited [21]. Thus, we hypothesize that the seedlings of woody species would have a lower N resorption than adults in a nutrient-poor environment.

We investigated N uptake and utilization including resorption in the drought-tolerant shrub *Calligonum caput-medusae* under different irrigation conditions during seedling establishment in the southern margin of the Taklimakan Desert, Xinjiang, China. Soils in this area are highly weathered and strongly leached, leading to very low nutrient concentrations [41].

C. caput-medusae is a perennial C4 plant belonging to Polygonaceae. Its foliage is reduced, making the assimilating green branches the primary photosynthetic organs. It continues to produce new green branches and shed old branches throughout the entire growing season. In addition, the older parts of the green branches of its seedlings become lignified, but more biomasses are allocated to the stem wood as the plant grows. We analyze how seedlings and 28-month-old well-established *C. caput-medusae* plants satisfy their N demand under different water regimes. We compare the effects of water addition on plant growth and N use characteristics to determine the relationships between water availability and N absorption as well as the utilization in seedling and established plants with senescent and green branches of the woody leafless shrub.

2. Materials and Methods

2.1. Study Area

The study was carried out in Cele Oasis which is located at the southern margin of the Taklimakan Desert in southern Xinjiang, China (Figure 1). The oasis has a typical arid continental climate with an annual mean temperature of 11.9 °C, mean annual precipitation of less than 40 mm (mainly occurring in May and July), and evaporation of approximately 2600 mm. Temperature ranges from 42 °C in summer to −24 °C in winter. The average annual wind speed is 1.9 $m \cdot s^{-1}$, and the maximum speeds in excess of 20.0 $m \cdot s^{-1}$ occur on more than 40 days per year. The frost-free period lasts 209 days per year. The soil is classified as aeolian sandy soil and irrigated desert soil according to the Chinese Soil Taxonomy; they are equivalent to Entisols and Inceptisols in the U.S. Soil Taxonomy, respectively. The Cele Oasis is surrounded by a 5 to 10 km belt of sparse vegetation (5–20% coverage) dominated by *Alhagi sparsifolia*, *Karelinia caspica* and *Tamarix ramosissima*.

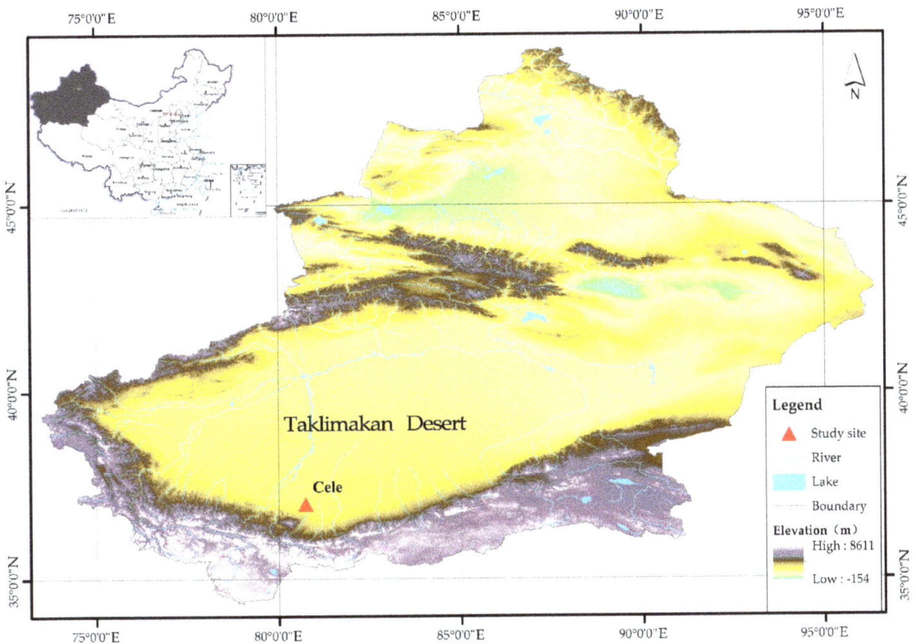

Figure 1. Study site at the southern margin of the Taklimakan Desert in southern Xinjiang, China.

2.2. Experimental Design

The pot experiment was initiated in April 2011 in an isolated and enclosed natural site to avoid disturbance. The experimental soil (0–40 cm) was collected from an oasis-desert ecotone, was air-dried, and passed through a 2 mm sieve. The characteristics of the experimental soil are shown in Table 1. A total of 85 kg soil was placed in each of the 150 plastic pots (40 cm inner diameter at the bottom, 50 cm inner diameter at the top edge, and 60 cm tall), to a bulk density of 1.4 g·cm^{-3}. The soil field capacity (FC) in the pots was 18%. Seven uniform holes had been drilled in the bottom of each pot and two layers of nylon mesh (0.25 mm) were placed over the holes to prevent root growth out of the pot, but aeration and drainage were. The pots were set in soil to provide thermal buffering with the top edge of each pot extending 3 cm above the ground. A plastic plate was placed on the bottom of each pot to eliminate water transfer into the surrounding soil. Healthy seeds of *C. caput-medusae* with similar sizes were collected in the autumn of 2010, and eight seeds were sown into each pot, and each seed was placed in 2 cm deep holes 5 cm apart on 5 April 2011. For the initial seedling establishment, all pots were well watered (soil moisture was approximately 80% FC) to ensure seed germination. When the one-month seedlings were thinned to one plant per pot, and different water treatments were then initiated. The shoots and roots of the 20 thinned seedlings were harvested and oven-dried at 75 °C, and then weighed to obtain the total biomass of the one-month-old seedlings with 2.4 g plant^{-1} on average.

Table 1. Basic soil characteristics before the start of the experiment.

pH	Organic Matter (g kg^{-1})	Total N (g kg^{-1})	Total P (g kg^{-1})	Total K (g kg^{-1})	Inorganic N (mg kg^{-1})	Available P (mg kg^{-1})	Available K (mg kg^{-1})
8.04	2.18	0.17	0.49	14.6	16.71	1.7	145.7

The pots were completely randomized and allocated to five water treatments (Figure 2), namely, water-stressed (30% and 40% FC), moderately-watered (50% and 60% FC), and well-watered (85% FC) with 30 replicates for each water level (10 per harvest). During the experimental period, volumetric soil water contents in four randomly selected replicate pots of each water level were measured using a moisture meter type TDR 300 (soil moisture equipment, Santa Barbara, CA, USA) every other day at 20:00. The amount of water to add to each pot was calculated according to the average of four measured pots under each water treatment and enabled the soil volumetric water content to be maintained at 5.4%, 7.2%, 9.0%, 10.8%, and 15.3%, and under 30%, 40%, 50%, 60%, and 85% FC water supply regimes, respectively. The water added to the plants was obtained from the local well without purification. We ignored the N effect from the well water because it contained little nitrate-N (2.92 mg·L^{-1}) and no ammonium-N. The water treatments were stopped at the end of the growing season (mid-November) and were re-started at the beginning of the next growing season (mid-April). The experiment was conducted for three years.

Figure 2. Experimental pots and growth status of *C. caput-medusae* after 2 months of water treatment.

2.3. Sample Collection

Plants were harvested three times from seedling to mature stages (Table 2). At each harvest time, four individual plants of each water treatment were randomly selected and separated to stem, branches, and roots. For each individual plant, all senesced but still attached branches were collected from each individual and combined into one sample per pot. All plant samples were oven-dried at 75 °C for 48 h and then weighed. Soil samples from 0–15, 15–30, and 30–45 cm depths in each pot were collected after the plant sampling. In each pot, three 2 cm-diameter soil cores were sampled by hand-auger and combined as a single composite sample. All fresh soil samples were sieved through a 2 mm mesh sieve to remove roots and stones and then divided into two subsamples.

Table 2. The sampling arrangement and parameters of analysis.

Sampling Time	Age	Samples	Parameters
Mid-August 2011 Late October 2011 Mid-August 2013	4 months 7 months 28 months	Green and senesced branches, stems, and roots; soil in each sampling pot	Biomass, total N concentration in each organ, soil total N, nitrate-N (NO_3^-), ammonium-N (NH_4^+) concentrations

2.4. Laboratory Analysis

All plant samples were ground with a ball mill and then analyzed for the total N concentration by the Kjeldahl acid-digestion method [42] with an Alpkem auto analyzer (Kjeltec System 8400 distilling unit, Foss, Copenhagen, Denmark). The N concentrations were expressed on a mass basis. One soil subsample (10 g) was freshly extracted with 50 mL of 0.01 M CaCl and analyzed for NO_3^- and NH_4^+ with a continuous flow analysis system (SEAL Analytical, Norderstedt, Germany). The other soil subsample was air-dried, passed through a 0.25 mm mesh, and then analyzed for soil characteristics. Soil total N concentration was determined by the semimicro Kjeldahl method [43]. Soil pH was determined with a 1:5 soil/water suspension; soil bulk density was measured by the soil core method; soil organic matter was determined by wet oxidation; the total phosphorus (P) was determined after digestion with spectrophotometer detection; soil-available P was extracted with 0.5 M $NaHCO_3$ solution and measured by colorimetric detection; soil total potassium (K) and available K were determined using a flame photometer [44].

2.5. Calculation

Individual plant biomass was calculated as the sum of the biomass for each organ (stem, branches, and roots). The total N uptake of each pot was calculated as the sum of individual organ N pools, where individual organ N pool was calculated by multiplying biomass (g·plant^{-1}) and its N concentration (mg·g^{-1}). The N use efficiency (NUE) was calculated as the ratio of total biomass to total N mass in the whole plant [45]. Nitrogen resorption efficiency (NRE) was calculated as NRE = $(1 - N_{senesced}/N_{green}) \times 100\%$, where $N_{senesced}$ and N_{green} are the N concentrations in senesced and green branches, respectively [19]. We used the reciprocal of $N_{senesced}$ to calculate N resorption proficiency (NRP), where a lower N concentration in senesced tissue corresponds to a higher proficiency [19].

Relative growth rate (RGR) was calculated as the increase in biomass over time: RGR = $(\log_{10}M_f - \log_{10}M_i)/(t_f - t_i)$, where M_i and M_f are the total individual biomass at the end of the fourth month and the first month, the seventh month and the fourth month, and the twenty-eighth month and the seventh month.

2.6. Statistical Analysis

One-way analysis of variance (ANOVA) was used to assess the effects of different water treatments on plant biomass, RGR, total N pool, NRE, NUE, NRP, N concentration, and soil N. Means were compared by Duncan's tests where ANOVA showed a significant difference. Two-way ANOVA was used to test the effects of plant age and water addition on soil N and plant N parameters. Regression analyses were used to determine the relationships among plant biomass, soil inorganic N, and plant N traits within each growth age. All the analyses were carried out with SPSS 16.0 (SPSS Inc., Chicago, IL, USA). We used the package "piecewise structural equation modeling" (piecewise SEM) to analyze the direct and indirect influences of water and plantation age on N concentrations in green and senesced branches, soil inorganic N, and plant biomass with R software (version 4.0.3) [46]. D-separation test of piecewise SEM was used to verify whether the causal model has important links, and $p > 0.05$ indicates the fitness of the model [47].

3. Results and Discussion

3.1. Soil N status

Average soil inorganic N concentration decreased from 16.71 mg·kg^{-1} to 3.43, 2.55 and 1.9 mg·kg^{-1} after water treatment for four months, seven months, and twenty-eight months, respectively (Figure 3a). Soil inorganic N concentrations under four- and seven-month treatments decreased with increasing water addition and reached the low values at 28 months. However, no relationship was found between water addition rate and soil inorganic N concentration (Table 3). This finding is consistent with several previous findings that soil N availability decreased as juvenile stands begin to mature [23,48]. However, soil total N concentration increased with plantation age from 0.09 g·kg^{-1} (initial value) to an average of 0.15 g·kg^{-1} (28 months) (Figure 3b). Water addition significantly increased soil total N concentration under 50% FC and 60% FC water treatments at 28 months, but no effects were found in four- and seven-month treatments. Therefore, some studies reported that the decline in soil inorganic N may be related to the slow release of available N through litter decomposition, N mineralization, and nitrification [49,50].

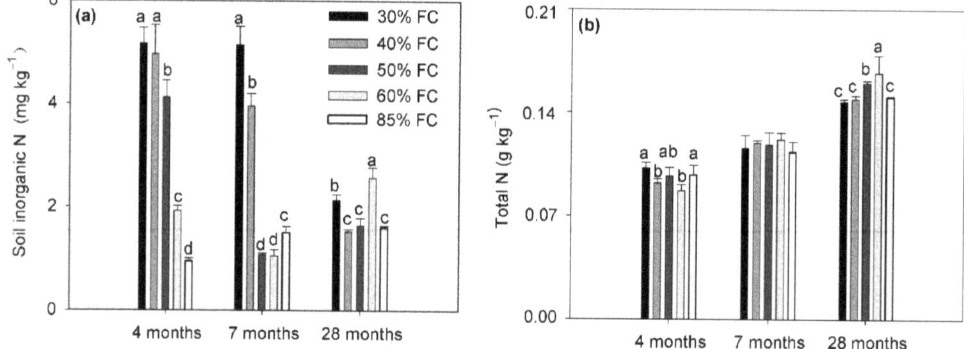

Figure 3. Soil inorganic N (**a**) and total N (**b**) concentrations in response to different water treatments at three growth stages (4 months, 7 months and 28months). Values are shown as the means ± se ($n = 4$). Bars with different lowercase letters indicate significant differences among treatments at the same growth stage at $p < 0.05$.

Table 3. Pearson's correlation coefficients between water addition rate and measured plant trait and soil N parameters for plants at 4-, 7-, and 28-month age.

Parameters	Water Addition Rate		
	4-Month	**7-Month**	**28-Month**
Soil inorganic N	−0.940 **	−0.735 **	−0.113
Soil total N	−0.136	−0.168	0.216
Individual biomass	0.947 **	0.984 **	0.995 **
Relative growth rate	0.964 **	0.890 **	0.597 **
Total N uptake	0.736 **	0.956 **	0.997 **
Green branch N concentration	−0.721 **	−0.638 **	0.362
Senesced branch N concentration	−0.939 **	−0.515 *	−0.685 **
N resorption efficiency	0.766 **	0.053	0.665 **
N use efficiency	0.491 *	0.514 *	−0.879 **

Notes: * $p < 0.05$; ** $p < 0.01$.

3.2. Plant Biomass and Relative Growth Rate

Individual plant biomass and RGR were significantly affected by plant age, water treatments, and their interaction (Table 4). The biomass increased significantly with ages and increasing water supply markedly enhanced this trend at three growth stages (Figure 4a), suggesting that *C. caput-medusae* showed strong adaptability to the decline in soil available N. This finding was consistent with the results in arid and semi-arid areas, indicating that water availability was positively correlated with productivity [30,51]. Increased water availability could directly stimulate plant physiological processes, and consequently increase net carbon uptake [51,52], resulting in high biomass accumulation. Nonetheless, the RGR decreased with plant age, at 0.35, 0.22, and 0.05 for four-, seven-, and 28-month-old saplings, respectively (Figure 4b). The RGR was also improved with an increasing water supply at three growth stages. However, the fourth- and seven-month-old plants grew much faster than the 28-month-old plants.

Table 4. Results (*F* value) of two-way ANOVA on the effects of water supply (W), plantation ages (A) and their interactions on soil N, plant biomass and plant N traits.

Treatments	$N_{inorganic}$	N_{total}	Biomass	RGR	N pool	N_{green}	$N_{senesced}$	NRE	NUE
Water (W)	289.7 ***	7.11 ***	1721 ***	96 ***	3570 ***	32.1 ***	43.2 ***	22.1 ***	38 ***
Age (A)	211.2 ***	233.3 ***	12,700 ***	3492 ***	17,740 ***	54.9 ***	168.6 ***	60 ***	1161 ***
W × A	117.3 ***	1.4	1427 ***	19 ***	2963 ***	34.5 ***	15.1 ***	10.5 ***	115.3 ***

Notes: $N_{inorganic}$ and N_{total} correspond to soil inorganic and total N concentrations; RGR and N pool correspond to relative growth rate and whole plant N pool; N_{green} and $N_{senesced}$ correspond to N concentrations in green and senesced branches; NRE, and NUE correspond to N resorption efficiency and N use efficiency; *** $p < 0.001$.

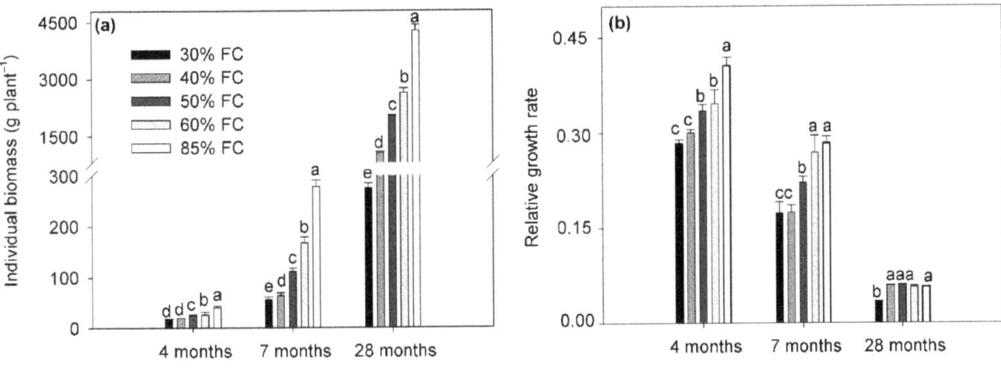

Figure 4. Changes in individual plant biomass (**a**) and relative growth rate (**b**) of *Calligonum caput-medusae* under different water treatments at three growth stages (4 months, 7 months, and 28 months). Values are shown as the means ± se (n = 4). Bars with different lowercase letters indicate significant differences among treatments at the same growth stage at $p < 0.05$.

3.3. Plant N Status

Water addition, plant age, and their interactions had significant effects on N concentrations in green and senesced branches (Table 4, Figure 5b,c). Plant age was the dominant factor in determining N concentrations. Average green branch N concentration among different water treatments decreased from 14.7 mg·g^{-1} at 4 months old to 11.5 mg·g^{-1} at 28 months old; these findings are lower than the average N concentration of terrestrial plant species (18.6 g·kg^{-1}) based on a global study [53]. Senesced branch N concentrations also decreased with plant age, with mean values of 5.3, 4.9, and 2.6 mg·g^{-1} at four-, seven- and 28-month-old, respectively. These results were lower than the critical value (7 mg·g^{-1}) reported by Killingbeck [19], implying that the senesced branch N was resorbed almost completely which leads to low litterfall N return to the soil. N was the limiting nutrient for

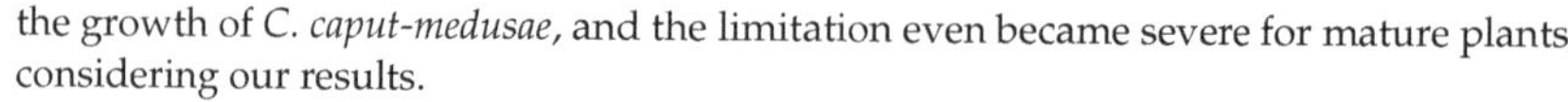

the growth of *C. caput-medusae*, and the limitation even became severe for mature plants, considering our results.

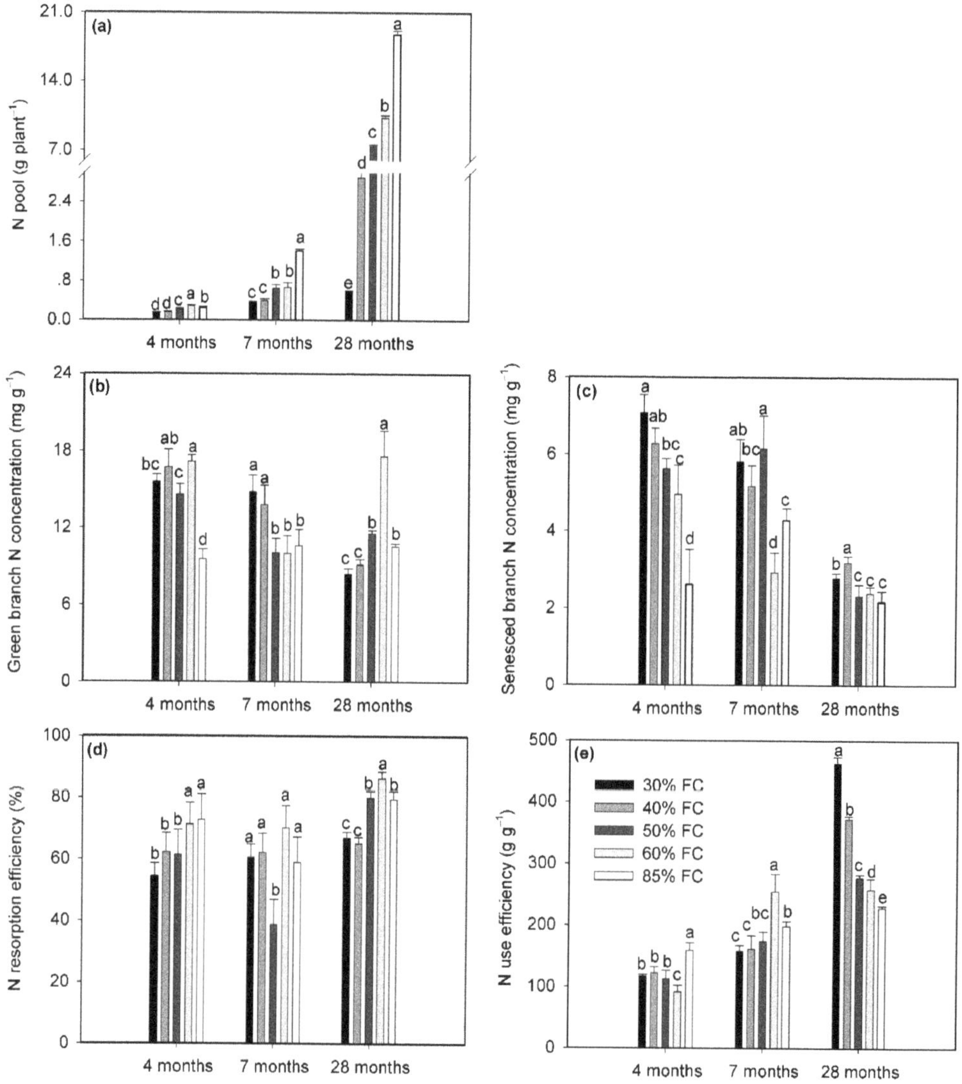

Figure 5. Nitrogen pools of total plant (**a**), N concentration in green (**b**) and senesced (**c**) branches, and N resorption efficiency (**d**) and N use efficiency (**e**) of *Calligonum caput-medusae* in response to different water treatments at three growth stages (4-month-old, 7-month-old, and 28-month-old). Values are shown as the means ± se (n = 4). Bars with different lowercase letters indicate significant differences among treatments at the same growth stage at $p < 0.05$.

Effects of water addition on green branch N concentration varied with plant ages. Increasing water addition significantly reduced green branch N concentrations of four- and seven-month-old plants but increased that of the 28-month-old plants. This finding is different from the observation in the shallow-rooted annuals and deep-rooted shrubs in the Gurbantunggut Desert, where water addition showed no effects on their green leaf N concentration [15]. Our study further found that plant N pool increased greatly with water addition rates (Figure 5a), suggesting that enhanced water supply significantly increased plant N uptake from soil [30]. According to these findings, we speculate that the increase in plant biomass may be higher than that of plant N pool with increasing water addition. This finding leads to the dilution effect of biomass on N content in green branches at the seedling stage, but it was inverse at the mature growth stage. This result was evidenced by the plant N pool at 28-month-age that responded more strongly to water addition than at four- and seven-month-age, showing an increase of 30.6 times at 85% FC water treatment relative to that at 30% FC water treatment. Different from green branches, N concentration in senesced branches was significantly and negatively correlated with water addition rates at all three plant growth ages. On the contrary, the NRP (the reciprocal of senesced branch N concentration) increased with increasing water addition. The decline in senesced branch N concentrations may lead to a decrease in N return to the soil.

3.4. N Resorption and Utilization

Non-parasitic plants have mainly two pathways for non-parasitic plants to acquire nutrients for new tissue production, as follows: root uptake from the soil, and mobilizing and withdrawing from old organs. The maintenance of N requirements for *C. caput-medusae* seedlings may be achieved through the pathways at the same time, as evidenced by the sharp decline of soil inorganic N and high NRE. The NRE of four- and seven-month-old plants was lower than that of 28-month-old plants, with mean values of 64.4%, 58.1%, and 75.5% (Figure 5d). The NRE of mature plants was higher than the global mean value (62%) [54]. Sun et al. [23] and Han et al. [55] suggested that plants mainly depended on N resorption with increasing limitation of soil available N. Thus, the mature *C. caput-medusae* may have changed its N acquisition process and depended less on root N uptake.

A significant positive relationship was found between NRE and water addition rates for four- and 28-month-old plants, but not for seven-month-old plants (Table 3), resulting in a significant interaction. In addition, the response of NRP to water addition was more significant than that of NRE, further confirming NRP. Thus, N levels in senesced branches were more sensitive for testing plant internal N cycling [27,56]. This finding suggested that enhanced soil water availability could improve plant's dependence on resorption-derived N due to the increasing limitation of soil available N. N use efficiency was also affected by water addition, plant age, and their interaction (Table 4). The NUE increased with plant age, with mean values of 120.5, 189.0, and 318.2 $g \cdot g^{-1}$ for four-, seven-, and 28-month-old plants, respectively (Figure 5e). Reports showed that plants could use limit N more efficiently with increasing water availability [57,58]. However, our results showed that water addition increased NUE of the four- and seven-month-old plants but decreased that of 28-month-old plants. This finding may be due to the mature plants that changed their N use strategies.

3.5. Controlling Factors of Plant Growth and N Utilization as Well as Corelationships between Them

Soil inorganic N concentration was positively correlated with green branch N concentrations at all three plant growth ages, but only related to senesced branch N concentrations at four months (Figure 6). Correspondingly, soil inorganic N was negatively correlated with NRE for four-month-old plants, but this finding was not found at other stages. This result may be related to the calculation of NRE based on the percent changes in green and senesced branch N concentrations. Our SEM result showed that green branch N concentrations were mainly determined by soil inorganic N concentration with indirect regulation by plant age and water addition (Figure 7), whereas soil inorganic N concentration and senesced branch N concentrations were mainly determined by water addition. The NRE was largely determined by senesced branch N concentrations, descendingly by green branch N and soil inorganic N concentrations. This finding suggested that the green branch N concentration could directly reflect the response of soil N availability to water addition during the establishment of *C. caput-medusae*. The plant NRE was not always closely related to the changes in green branch N concentrations and soil N availability. The indirect impact of water addition (via changes in senesced branch N concentrations) is an important driver that changes the N resorption of *C. caput-medusae*.

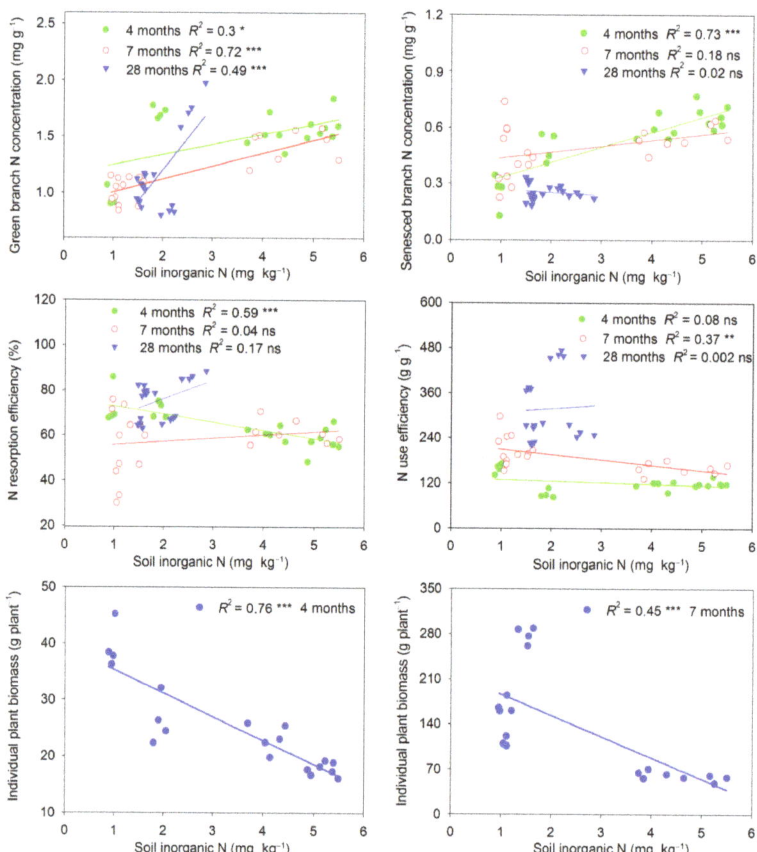

Figure 6. Relationships between N concentrations in green and senesced branches, N resorption efficiency, N use efficiency, individual plant biomass, and soil inorganic N. There was no correlation between soil inorganic N and individual plant biomass at 28 months old (R^2 = 0.01), and thus not displayed in the plots. Note: * p < 0.05; ** p < 0.01; *** p < 0.001; ns indicates no significant.

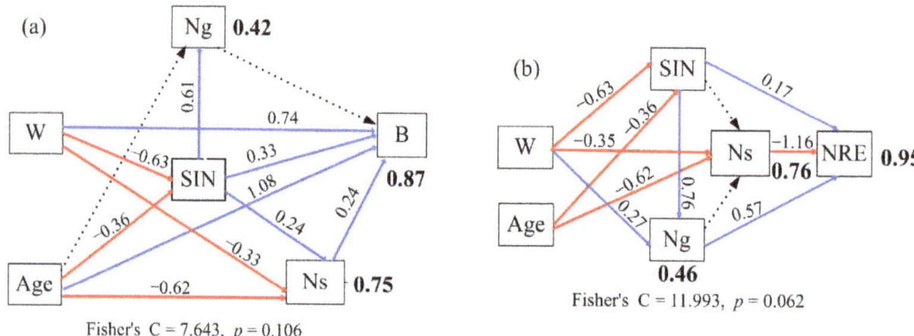

Figure 7. Controlling factor analysis of N resorption efficiency and individual plant biomass using the structural equation model. Solid and dashed lines indicate significant (p < 0.05) and non-significant (p > 0.05) regressions. Blue and red arrows represent positive and negative relationships, respectively. W, water; B, individual plant biomass; SIN, soil inorganic N concentration; Ng, N concentrations in green branches; Ns, N concentrations in senesced branches; NRE, N resorption efficiency. (**a**) controlling factor analysis of individual plant biomass; (**b**) controlling factor analysis of N resorption efficiency.

Individual plant biomass was significantly and negatively correlated with soil inorganic N and green branch N concentrations for the four- and seven-month-old plants, but it was not found for 28-month-old plants (Figure 8). Senesced branch N concentration was strongly and negatively related with individual plant biomass at all three measured times. Therefore, the plant biomass was directly affected by senesced branch N and soil inorganic N concentrations (Figure 7). Our SEM result also showed that water addition played an important direct role in driving plant biomass during the period of seedling establishment of *C. caput-medusae*. The water supply could—both directly and indirectly—regulate plant growth at the same time.

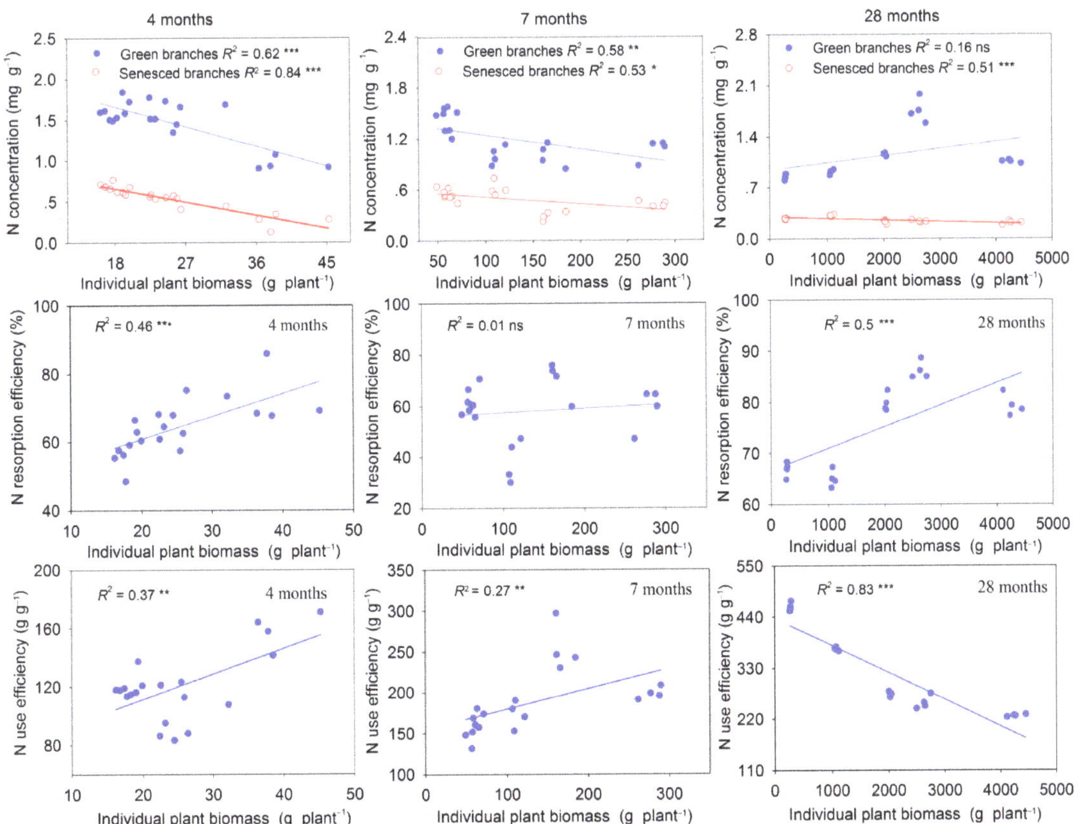

Figure 8. Relationships between N concentration in green and senesced branches, N resorption efficiency, N use efficiency, and individual plant biomass of *Calligonum caput-medusae*. Note: * $p < 0.05$; ** $p < 0.01$; *** $p < 0.001$; ns indicates no significant.

Accordingly, we found that plant biomass was positively related to NUE for young plants, but negatively related between them for mature plants (Figure 8). However, plant biomass was positively correlated with NRE at four and 28 months but positively related to NRP at three growth stages. This result indicated that plant biomass and N acquisition were maintained by increasing NUE and N resorption at the seedling stage. Water addition also showed positive effects on these processes. Thus, this result also provided evidence for our finding that plant N acquirement depended on soil N and resorbed N for the saplings. The increasing biomass production of mature plants may be closely related to the high N resorption ability, which may explain the decrease in NUE with water supply. This finding was supported by previous studies, where plants with high productivity had high N resorption [23,59]; thus, their dependence on soil-available N supply was reduced.

4. Conclusions

Our results showed that the limitation of soil available N became increasingly serious with plant growth and was exacerbated by water addition. It appears to be an N depletion process for the seedling establishment of *C. caput-medusae*. However, the plants showed strong adaptability to N limitation and satisfied their N requirement by increasing plant N pool, NUE, and N resorption at the seedling stage, but mainly depended on N resorption at the mature stage. Our SEM showed that the individual plant biomass was largely determined by plant age and water addition, and subsequently by soil inorganic N and senesced branch N concentrations which were regulated by plant age and water addition. Enhanced water supply significantly improved plant N uptake from soil and negatively affected soil available N. Water addition mainly promoted NRE by reducing senesced branch N concentrations to maintain plant productivity over the period of seedling establishment of *C. caput-medusae*. Our findings provide a better insight to understand the N adaptive responses to irrigation and lay the groundwork for the vegetation establishment in the hyper-arid ecosystem. Future research is required to explore whether the resorption-derived process can satisfy plant N requirement for a longer time in the study area.

Author Contributions: Writing—original draft preparation, C.H.; writing—review and editing, C.H., F.Z. and S.Z.; supervision, S.Z.; project administration, C.H. and F.Z.; methodology, S.Z. and B.Z.; formal analysis, J.X.; software and analysis, S.Z. and J.X. All authors have read and agreed to the published version of the manuscript.

Funding: This research was funded by Key Laboratory Project of Xinjiang Uygur Autonomous Region (E0310113), the Original Innovation Project of the Basic Frontier Scientific Research Program, Chinese Academy of Sciences (ZDBS-LY-DQC031), the National Natural Science Foundation of China (42071259), and Natural Science Foundation of Xinjiang Uygur Autonomous Region (2021D01E01).

Institutional Review Board Statement: Not applicable.

Informed Consent Statement: Not applicable.

Data Availability Statement: All data reported here is available from the authors upon request.

Acknowledgments: We would like to thank Jonathan R Leake for his helpful suggestions to improve the manuscript.

Conflicts of Interest: The authors declare no conflict of interest.

References

1. Klein Goldewijk, K.; Beusen, A.; Doelman, J.; Stehfest, E. Anthropogenic land use estimates for the Holocene—HYDE 3.2. *Earth Syst. Sci. Data* **2017**, *9*, 927–953. [CrossRef]
2. Hu, Y.L.; Zeng, D.H.; Fan, Z.P.; Chen, G.S.; Zhao, Q.; Pepper, D. Changes in ecosystem carbon stocks following grassland afforestation of semiarid sandy soil in the southeastern Keerqin Sandy Lands, China. *J. Arid Environ.* **2008**, *72*, 2193–2200. [CrossRef]
3. Lei, J.Q.; Li, S.Y.; Jin, Z.Z.; Fan, J.L.; Wang, H.F. Comprehensive eco-environmental effects of the shelter-forest ecological engineering along the Tarim Desert Highway. *Chin. Sci. Bull.* **2008**, *53*, 190–202. [CrossRef]
4. Wei, W.; Chen, L.D.; Yang, L.; Samadani, F.F.; Sun, G. Microtopography recreation benefits ecosystem restoration. *Environ. Sci. Technol.* **2012**, *46*, 10875–10876. [CrossRef]
5. Xu, X.W.; Li, B.W.; Wang, X.J. Progress in study on irrigation practice with saline groundwater on sandlands of Taklimakan Desert hinterland. *Chin. Sci. Bull.* **2006**, *51*, 161–166. [CrossRef]
6. Shan, L.S.; Li, Y.; Zhao, R.F.; Zhang, X.M. Effects of deficit irrigation on daily and seasonal variations of trunk sap flow and its growth in *Calligonum arborescens*. *J. Arid Land* **2013**, *5*, 233–243. [CrossRef]
7. Li, C.; Shi, X.; Mohamad, O.A.; Gao, J.; Xu, X.; Xie, Y. Moderate irrigation intervals facilitate establishment of two desert shrubs in the Taklimakan Desert Highway Shelterbelt in China. *PLoS ONE* **2017**, *12*, e0180875. [CrossRef]
8. Hooper, D.U.; Johnson, L. Nitrgen limitation in dryland ecosystems: Responses to geographical and temporal variation in precipitation. *Biogeochemistry* **1999**, *46*, 247–293. [CrossRef]
9. Misra, A.; Tyler, G. Effects of wet and dry cycles in calcareous soil on mineral nutrient uptake of two grasses, *Agrostis stolonifera* L. and *Festuca ovina* L. *Plant Soil* **2000**, *224*, 297–303. [CrossRef]

10. Wang, C.H.; Wan, S.Q.; Xing, X.R.; Zhang, L.; Han, X. Temperature and soil moisture interactively affected soil net N mineralization in temperate grassland in Northern China. *Soil Biol. Biochem.* **2006**, *38*, 1101–1110. [CrossRef]
11. Schlesinger, W.H.; Dietze, M.C.; Jackson, R.B.; Phillips, R.P.; Rhoades, C.C.; Rustad, L.E.; Vose, J.M. Forest biogeochemistry in response to drought. *Glob. Chang. Biol.* **2016**, *22*, 2318–2328. [CrossRef] [PubMed]
12. Zhou, X.B.; Bowker, M.A.; Tao, Y.; Wu, L.; Zhang, Y.M. Chronic nitrogen addition induces a cascade of plant community responses with both seasonal and progressive dynamics. *Sci. Total Environ.* **2018**, *626*, 99–108. [CrossRef] [PubMed]
13. Shi, W.H.; Lin, L.; Shao, S.L.; He, A.G.; Ying, Y.Q. Effects of simulated nitrogen deposition on *Phyllostachys edulis* (Carr.) seedlings under different watering conditions: Is seedling drought tolerance related to nitrogen metabolism? *Plant Soil* **2020**, *448*, 539–552. [CrossRef]
14. Zhou, X.B.; Zhang, Y.M.; Niklas, K.J. Sensitivity of growth and biomass allocation patterns to increasing nitrogen: A comparison between ephemerals and annuals in the Gurbantunggut Desert, north-western China. *Ann. Bot.* **2014**, *113*, 501–511. [CrossRef]
15. Huang, G.; Su, Y.G.; Mu, X.H.; Li, Y. Foliar nutrient resorption responses of three life-form plants to water and nitrogen additions in a temperate desert. *Plant Soil* **2018**, *424*, 479–489. [CrossRef]
16. Li, K.H.; Liu, X.J.; Geng, F.Z.; Xu, W.; Lv, J.L.; Dore, A.J. Inorganic nitrogen deposition in arid land ecosystems of Central Asia. *Environ. Sci. Pollut. Res.* **2021**, *28*, 31861–31871. [CrossRef] [PubMed]
17. Killingbeck, K.T. The terminological jungle revisited: Making a case for use of the term resorption. *Oikos* **1986**, *46*, 263–264. [CrossRef]
18. Reich, P.B.; Walters, M.B.; Ellsworth, D.S. Leaf life-span in relation to leaf, plant, and stand characteristics among diverse ecosystems. *Ecol. Monogr.* **1992**, *62*, 365–392. [CrossRef]
19. Killingbeck, K.T. Nutrients in senesced leaves: Keys to the search for potential resorption and resorption proficiency. *Ecology* **1996**, *77*, 1716–1727. [CrossRef]
20. Aerts, R.; Chapin, F.S., III. The mineral nutrition of wild plants revisited: A re-evaluation of processes and patterns. *Adv. Ecol. Res.* **1999**, *30*, 1–67.
21. Wright, I.J.; Westoby, M. Nutrient concentration, resorption and lifespan: Leaf traits of Australian sclerophyll species. *Funct. Ecol.* **2003**, *17*, 10–19. [CrossRef]
22. Cleveland, C.C.; Houlton, B.Z.; Smith, W.K.; Marklein, A.R.; Running, S.W. Patterns of new versus recycled primary production in the terrestrial biosphere. *Proc. Natl. Acad. Sci. USA* **2013**, *110*, 12733–12737. [CrossRef] [PubMed]
23. Sun, Z.Z.; Liu, L.L.; Peng, S.S.; Peñuelas, J.; Zeng, H.; Piao, S.L. Age-related modulation of the nitrogen resorption efficiency response to growth requirements and soil nitrogen availability in a temperate pine plantation. *Ecosystems* **2016**, *19*, 698–709. [CrossRef]
24. Liu, P.; Huang, J.H.; Sun, O.J.; Han, X. Litter decomposition and nutrient release as affected by soil nitrogen availability and litter quality in a semiarid grassland ecosystem. *Oecologia* **2010**, *162*, 771–780. [CrossRef]
25. Suseela, V.; Tharayil, N.; Xing, B.S.; Dukes, J.S. Warming and drought differentially influence the production and resorption of elemental and metabolic nitrogen pools in *Quercus rubra*. *Glob. Chang. Biol.* **2015**, *21*, 4177–4195. [CrossRef] [PubMed]
26. Yan, T.; Lü, X.T.; Yang, K.; Zhu, J. Leaf nutrient dynamics and nutrient resorption: A comparison between larch plantations and adjacent secondary forests in Northeast China. *J. Plant Ecol.* **2016**, *9*, 165–173. [CrossRef]
27. Yuan, Z.Y.; Chen, H.Y.H. Negative effects of fertilization on plant nutrient resorption. *Ecology* **2015**, *96*, 373–380. [CrossRef] [PubMed]
28. Brant, A.N.; Chen, H.Y.H. Patterns and mechanisms of nutrient resorption in plants. *Crit. Rev. Plant Sci.* **2015**, *34*, 471–486. [CrossRef]
29. Yuan, Z.Y.; Chen, H.Y.H. Global-scale patterns of nutrient resorption associated with latitude, temperature and precipitation. *Glob. Ecol. Biogeogr.* **2009**, *18*, 11–18. [CrossRef]
30. Ren, H.Y.; Xu, Z.W.; Huang, J.H.; Lü, X.T.; Zeng, D.H.; Yuan, Z.Y.; Han, X.G.; Fang, Y.T. Increased precipitation induces a positive plant-soil feedback in a semi-arid grassland. *Plant Soil* **2015**, *389*, 211–223. [CrossRef]
31. Rivero, R.M.; Kojima, M.; Gepstein, A.; Sakakibara, H.; Mittler, R.; Gepstein, S.; Blumwald, E. Delayed leaf senescence induces extreme drought tolerance in a flowering plant. *Proc. Natl. Acad. Sci. USA* **2007**, *104*, 19631–19636. [CrossRef] [PubMed]
32. Lajtha, K. Nutrient reabsorption efficiency and the response to phosphorus fertilization in the desert shrub *Larrea-tridentata* (Dc) Cov. *Biogeochemistry* **1987**, *4*, 265–276. [CrossRef]
33. Huang, J.Y.; Yu, H.L.; Li, L.H.; Yuan, Z.Y.; Bartels, S. Water supply changes N and P conservation in a perennial grass *Leymuschinensis*. *J. Integr. Plant Biol.* **2009**, *51*, 1050–1056. [CrossRef] [PubMed]
34. Khasanova, A.; James, J.J.; Drenovsky, R.E. Impacts of drought on plant water relations and nitrogen nutrition in dryland perennial grasses. *Plant Soil* **2013**, *372*, 541–552. [CrossRef]
35. Ruehr, N.K.; Offermann, C.A.; Gessler, A.; Winkler, J.B.; Ferrio, J.P.; Buchmann, N.; Barnard, R.L. Drought effects on allocation of recent carbon: From beech leaves to soil CO_2 efflux. *New Phytol.* **2009**, *184*, 950–961. [CrossRef] [PubMed]
36. Vickers, A.D.; Palmer, S.C.F. The influence of canopy cover and other factors upon the regeneration of Scots pine and its associated ground flora within Glen Tanar National Nature Reserve. *Forestry* **2000**, *73*, 37–49. [CrossRef]
37. Zhang, J.G.; Lei, J.Q.; Wang, Y.D.; Zhao, Y.; Xu, X.W. Survival and growth of three afforestation species under high saline drip irrigation in the Taklimakan Desert, China. *Ecosphere* **2016**, *7*, e01285. [CrossRef]

38. Nong, H.T.; Tateishi, R.; Suriyasak, C.; Kobayashi, T.; Oyama, Y.; Chen, W.J.; Matsumoto, R.; Hamaoka, N.; Iwaya-Inoue, M.; Ishibashi, Y. Effect of seedling nitrogen condition on subsequent vegetative growth stages and its relationship to the expression of nitrogen transporter genes in rice. *Plants* **2020**, *9*, 861. [CrossRef] [PubMed]
39. Panda, M.M.; Reddy, M.D.; Sharma, A.R. Yield performance of rainfed lowland rice as affected by nursery fertilization under conditions of intermediate deepwater (15–50 cm) and flash floods. *Plant Soil* **1991**, *132*, 65–71. [CrossRef]
40. Bond-Lamberty, B.; Gower, S.T.; Wang, C.K.; Cyr, P.; Veldhuis, H. Nitrogen dynamics of a boreal black spruce wildfire chronosequence. *Biogeochemistry* **2006**, *81*, 1–16. [CrossRef]
41. Gui, D.W.; Lei, J.Q.; Zeng, F.J. Farmland management effects on the quality of surface soil during oasification in the southern rim of the Tarim Basin in Xinjiang, China. *Plant Soil Environ.* **2010**, *56*, 348–356. [CrossRef]
42. Dordas, C.; Sioulas, C. Dry matter and nitrogen accumulation, partitioning, and retranslocation in safflower (*Carthamus tinctorius* L.) as affected by nitrogen fertilization. *Field Crops Res.* **2009**, *110*, 35–43. [CrossRef]
43. Bremner, J.M.; Mulvaney, C.S. Nitrogen-total. In *Methods of Soil Analysis*, 2nd ed.; Page, A.L., Miller, R.H., Keeney, D.R., Eds.; American Society of Agronomy: Madison, WI, USA, 1982; pp. 595–624.
44. Carter, M.R. *Soil Sampling and Methods of Analysis*; Lewis Publishers CRC Press: Boca Raton, FL, USA, 1993.
45. Shaver, G.R.; Melillo, J.M. Nutrient budgets of marsh plants: Efficiency concepts and relation to availability. *Ecology* **1984**, *65*, 1491–1510. [CrossRef]
46. Lefcheck, J.S. Piecewise SEM: Piecewise structural equation modeling in R for ecology, evolution, and systematics. *Methods Ecol. Evol.* **2016**, *7*, 573–579. [CrossRef]
47. Shipley, B. *Cause and Correlation in Biology: A User's Guide to Path Analysis, Structural Equations and Causal Inference*; Cambridge University Press: Cambridge, UK, 2002.
48. Tang, J.W.; Luyssaert, S.; Richardson, A.D.; Kutsch, W.; Janssens, I.A. Steeper declines in forest photosynthesis than respiration explain age-driven decreases in forest growth. *Proc. Natl. Acad. Sci. USA* **2014**, *111*, 8856–8860. [CrossRef] [PubMed]
49. Aerts, R. Climate, leaf litter chemistry and leaf litter decomposition in terrestrial ecosystems: A triangular relationship. *Oikos* **1997**, *79*, 439–449. [CrossRef]
50. Farley, K.A.; Kelly, E.F. Effects of afforestation of a páramo grassland on soil nutrient status. *For. Ecol. Manag.* **2004**, *195*, 281–290. [CrossRef]
51. Wu, Z.T.; Dijkstra, P.; Koch, G.W.; Peñuelas, J.; Hungate, B.A. Responses of terrestrial ecosystems to temperature and precipitation change: A meta-analysis of experimental manipulation. *Glob. Chang. Biol.* **2011**, *17*, 927–942. [CrossRef]
52. Patrick, L.; Cable, J.; Potts, D.; Ignace, D.; Barron-Gafford, G.; Griffith, A.; Alpert, H.; Van Gestel, N.; Robertson, T.; Huxman, T.E.; et al. Effects of an increase in summer precipitation on leaf, soil, and ecosystem fluxes of CO_2 and H_2O in a sotol grassland in Big Bend National Park, Texas. *Oecologia* **2007**, *151*, 704–718. [CrossRef] [PubMed]
53. Han, W.X.; Fang, J.Y.; Guo, D.L.; Zhang, Y. Leaf nitrogen and phosphorus stoichiometry across 753 terrestrial plant species in China. *New Phytol.* **2005**, *168*, 377–385. [CrossRef] [PubMed]
54. Vergutz, L.; Manzoni, S.; Porporato, A. Global resorption efficiencies and concentrations of carbon and nutrients in leaves of terrestrial plants. *Ecol. Monogr.* **2012**, *82*, 205–220. [CrossRef]
55. Han, W.X.; Tang, L.Y.; Chen, Y.H.; Fang, J.Y. Relationship between the relative limitation and resorption efficiency of nitrogen vs phosphorus in woody plants. *PLoS ONE* **2013**, *8*, e83366. [CrossRef]
56. Rejmankova, E. Nutrient resorption in wetland macrophytes: Comparison across several regions of different nutrient status. *New Phytol.* **2005**, *167*, 471–482. [CrossRef] [PubMed]
57. Kost, J.A.; Boerner, R.E.J. Foliar nutrient dynamics and nutrient use efficiency in *Cornus florida*. *Oecologia* **1985**, *66*, 602–606. [CrossRef]
58. Li, J.Z.; Lin, S.; Taube, F.; Pan, Q.; Dittert, K. Above and belowground net primary productivity of grassland influenced by supplemental water and nitrogen in Inner Mongolia. *Plant Soil* **2011**, *340*, 253–264. [CrossRef]
59. Nambiar, E.K.S.; Fife, D.N. Nutrient retranslocation in temperate conifers. *Tree Physiol.* **1991**, *9*, 185–207. [CrossRef] [PubMed]

Article

Study on Dynamic Changes of Soil Erosion in the North and South Mountains of Lanzhou

Hua Zhang [1,2,*], Jinping Lei [1], Hao Wang [3], Cungang Xu [1] and Yuxin Yin [1]

1 College of Geography and Environmental Sciences, Northwest Normal University, Lanzhou 730070, China; 2020212672@nwnu.edu.cn (J.L.); cungangxu@163.com (C.X.); 2020212680@nwnu.edu.cn (Y.Y.)
2 Key Laboratory of Resource Environment and Sustainable Development of Oasis, Lanzhou 730070, China
3 Faculty of Geographic Science, Beijing Normal University, Beijing 100875, China; 18772102861@163.com
* Correspondence: zhanghua@nwnu.edu.cn

Abstract: The North and South Mountains of Lanzhou City are the ecological protection barriers and an important part of the ecological system of Lanzhou City. This study takes the North and South Mountains as the study area, calculates the soil erosion modulus of the North and South Mountains of Lanzhou City based on the five major soil erosion factors in the RUSLE model, and analyses the spatial and temporal dynamics of soil erosion in the North and South Mountains of Lanzhou City and the soil erosion characteristics under different environmental factors. The results of the study show that: The intensity of soil erosion is dominated by slight erosion, which was distributed in the northwestern and southeastern parts of the North and South Mountains in 1995, 2000, 2005, 2010, 2015 and 2018. Under different environmental factors, the soil erosion modulus increased with elevation and then decreased; the soil erosion modulus increased with a slope; the average soil erosion modulus of grassland was the largest, followed by forest land, cultivated land, unused land, construction land, and it was the smallest for water; except for bare land, the average soil erosion modulus decreases with the increase of vegetation cover; Soil erosion modulus was the greatest in the pedocal of the North and South Mountains, and the least in the alpine soil.

Keywords: soil erosion; RUSLE model; erosion intensity; land desertification; North and South Mountains of Lanzhou

Citation: Zhang, H.; Lei, J.; Wang, H.; Xu, C.; Yin, Y. Study on Dynamic Changes of Soil Erosion in the North and South Mountains of Lanzhou. *Water* **2022**, *14*, 2388. https://doi.org/10.3390/w14152388

Academic Editors: Ying Zhao, Jianguo Zhang, Jianhua Si, Jie Xue and Zhongju Meng

Received: 29 June 2022
Accepted: 29 July 2022
Published: 1 August 2022

1. Introduction

Soil erosion is the destruction and loss of soil and water resources and land productivity due to natural forces and human activities, mainly including land surface erosion and water loss, which is the most dominant form of soil degradation [1–3]. Soil erosion will destroy the surface structure, reduce land fertility, raise the riverbed, destroy water conservancy facilities, aggravate flood and drought, and pose a significant threat to agricultural production, river water quality, and the environment. Soil erosion has become one of the world's most extensive and complicated ecological problems. It has become the concern of many disciplines [4], such as soil science, agronomy, hydrology, environmental science, and so on [5–8]. China is one of the countries with the most severe soil erosion [9,10]. In 2018, the soil erosion area in China reached 2.73×10^6 km^2, accounting for about 28.80% of the total area in China except Taiwan Province, a large area and a wide distribution [11]. The area of soil erosion in Northwest China is 1.26×10^6 km^2, accounting for 40.95% of the total area of Northwest China, and this presents a fundamental environmental problem [12]. In Gansu Province, for instance, the soil erosion area reached 1.86×10^5 km^2, accounting for about 40.66% of the total area, which exerted significant pressure on soil and water conservation and the construction of ecological civilization.

Lanzhou City, the capital of Gansu Province, is located in the upper basin of the Yellow River and consists of a pearl-shaped basin formed by the alluvial deposits of the Yellow River. To the north and south of Lanzhou City are the North and South Mountains, a mountain range covered with loess formed by the terraces of the Yellow River. To the north of Lanzhou City is the Tengger Desert; to the west is the Badanjilin Desert, a region with 2.78×10 km^2 of severely desertified land [13], resulting in frequent sandstorms, serious surface exposure and fragile ecosystems [14–16]. The natural vegetation in the North and South Mountains of Lanzhou is mainly desert vegetation, at present, the area of the environmental greening project in the North and South Mountains of Lanzhou City has reached 413 km^2, with 1.6×10^8 trees of various types being established, forming a complete artificial ecosystem and making the North and South Mountains an important ecological barrier [17,18]. Therefore, it is of great significance for soil and water conservation and ecological civilization construction in Lanzhou to reveal the temporal and spatial characteristics of soil erosion and analyze the dynamic changes of soil erosion in Lanzhou.

Soil erosion models are a common method for quantitative soil erosion estimation. USDA and university cooperations saw the establishment of the Revised Universal Soil Loss Equation (RUSLE), based on the Universal Soil Loss Equation (USLE) in 1986 [19], which has become a widely used model for quantitative soil erosion estimation worldwide due to its simplicity, few parameter requirements, and high estimation accuracy compared to other soil erosion models [20–23]. Therefore, this study takes the South and North Mountains of Lanzhou City as the research area, takes soil erosion as the research content, and uses the RULSE model to calculate the soil erosion modulus of the South and North Mountains of Lanzhou City based on soil field sampling data, land use, and precipitation data. The objectives of this study are (1) to reveal the spatial and temporal variation characteristics of soil erosion in the South and North Mountains of Lanzhou City, and (2) to provide scientific reference for the construction of water and soil conservation and ecological civilization.

2. Materials and Methods

2.1. General Situations of the Study Area

The North and South Mountains of Lanzhou span Anning District, Qilihe District, Chengguan District, Xigu District, Gaolan County, and Yuzhong County within the jurisdiction of Lanzhou City, with geographical coordinates of 35°44′–36°19′ N, 103°21′–103°59′ E. The total area is about 1940.08 km^2 (Figure 1). Among them, the green part accounts for 846.66 km^2, and the non-green region accounts for 1147.42 km^2. The geological conditions of this area are involved. The topography is fragmented, and natural disasters are to occur easily. The climate type belongs to the temperate semi-arid continental monsoon climate, with an annual average temperature of 9.1 °C and a yearly average rainfall of 327.7 mm, mostly concentrated from July to September, and the average annual potential evaporation is 1468 mm, which is 4.4 times of the precipitation. The vegetation type belongs to the transition type of typical steppe to desert steppe. Currently, most of the existing forests in Lanzhou's North and South Mountains are artificial forests, mostly young and middle-aged. Artificial afforestation is mainly coniferous and broad-leaved mixed forest, arbor shrub mixed forest, and shrub forest. The soil types in this area are primarily grey calcareous soil, mostly dark grey calcareous soil and typically grey calcareous soil in the South Mountains, and light grey calcareous soil and red sandy soil in the northern mountain, with a loose texture and weak anti-erosion ability.

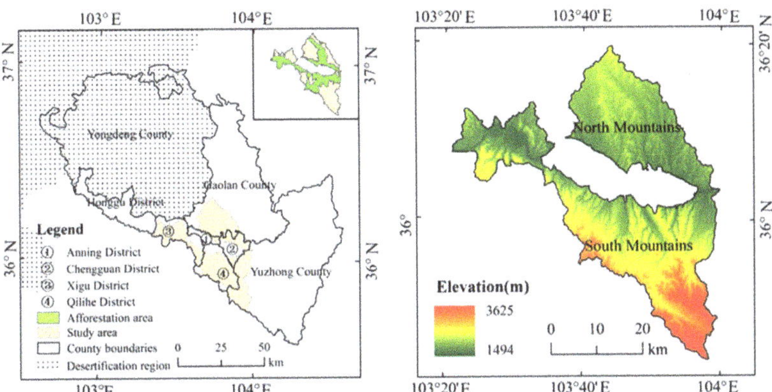

Figure 1. Overview of the study area.

2.2. Data Source

2.2.1. Soil Texture and Organic Carbon Data

(1) Soil sample sampling

Using the 1:1 million soil map of the North and South Mountains of Lanzhou City as the base map, about 120 soil sampling points were designed in July–August 2019 according to a uniform distribution method of 4 km × 4 km, The sampling was carried out according to the plan, combining the actual situation with randomly selected 10 m × 10 m sample plots. A total of 130 soil samples were collected. The field sampling operation was carried out using a soil auger to collect soil samples from 0–20 cm of the surface layer at the center of the sample plots and four right-angle points, mixed evenly and placed in self-sealing bags for the determination of soil texture and soil organic carbon. 0–20 cm soil samples of the surface layer at the center and four right corners were collected with a ring knife, put into an aluminum box, and weighed fresh at the sampling site, which was used to determine the soil bulk density. Using GPS positioning, the elevation, longitude, and latitude of the sampling points in the center were recorded and numbered sequentially.

(2) Determination of soil samples

The determination of soil texture was carried out by the Mastersizer 2000 laser particle size analyzer (model: MS2000, Zhenxiang Technology Co., Ltd., Changsha, China). The soil organic carbon content was determined by the Qiulin method, the soil salinity was determined by the "residue drying-mass method", and the pH value was determined by the "potential method".

2.2.2. Other Data

(1) The meteorological data was based on the monthly precipitation data set of 0.5° × 0.5° in China from 1995 to 2018 (V2.0), which came from the China Meteorological data sharing Network (http://data.cma.cn/), (accessed on 25 March 2019). (2) The GDEM-DEM 30 m spatial resolution digital elevation was derived from the geospatial data cloud (http://www.gscloud.cn/), (accessed on 25 March 2019). (3) The Landsat TM/OLI image from 1995 to 2018 was selected as the source of the Google Earth Engine cloud platform, to calculate the Normalized Difference Vegetation Index (NDVI) for the study area (Google Earth Engine, GEE) (https://earthengine.google.com/), (accessed on 25 March 2019). The image was programmed in the platform to preprocess the image. (4) The land use with a spatial resolution of 30 m in 1990, 2000, 2005, 2010, 2015, and 2018 was selected from the Resource and Environmental Science Data Center of the Chinese Academy of Sciences (http://www.resdc.cn/), (accessed on 25 March 2019). (5) National Cryosphere Desert Data

Center provided the desertification distribution data (http://www.ncdc.ac.cn), (accessed on 25 March 2019).

2.3. Research Methods

2.3.1. Soil Erosion Model

The study used the RUSLE model to estimate soil erosion in Lanzhou [24–26]. The formula is as follows:

$$A = R \cdot K \cdot LS \cdot C \cdot P \tag{1}$$

Among them, A is the average soil erosion amount per unit area last year, the unit is [t/(km^2·a)],and R is the precipitation erosivity factor, the unit is [MJ mm/(km^2·h·a)], K is the soil erodibility factor, in units [t·km^2·h/(km^2·MJ·mm)], LS is the slope length factor (dimensionless); C is the vegetation cover and management factor, (dimensionless); P is the soil and water conservation and measure factor (dimensionless).

2.3.2. Determination of Factors in the RUSLE Model

(1) Determination of R-value of precipitation erosivity factor

Precipitation is one of the important exogenous forces causing soil erosion, reflecting the potential impact of annual average or maximum precipitation on soil erosion. This study adopted the method of estimating rainfall erosivity by using yearly and monthly precipitation data proposed by Wischmeier [27]. The formula is as follows:

$$R = \sum_{i=1}^{12}\left[1.735 \times 10^{\left(1.5 \times \log \frac{P_i^2}{P} - 0.8188\right)}\right] \tag{2}$$

In the formula, P_i is monthly precipitation (mm); P is annual precipitation (mm). This method has been applied in the western region, and good results have been obtained km [28].

(2) K value of soil erodibility factor

The soil erodibility factor refers to the soil loss rate under a given unit of precipitation erosivity measured in a standard plot [29,30]. In this study, Williams' calculation method of soil erodibility factor K in the EPIC model was adopted [31]. The formula is as follows:

$$\begin{aligned} K = \; & 0.1317 \times \left\{0.2 + 0.3 \exp\left[-0.0256\, sand\left(1 - \frac{slit}{100}\right)\right]\right\} \times \left[\frac{silt}{clay + silt}\right]^{0.3} \\ & \times \left[1 - \frac{0.25C}{C + \exp(3.72 - 2.95C)}\right] \times \left[1 - \frac{0.7sn1}{sn1 + \exp(-5.51 + 22.9\, sn1)}\right] \end{aligned} \tag{3}$$

Among them, *Sand*, *Silt*, and *Clay* represent the percentage of sand, silt, and clay content in soil, respectively (%); C is the percentage of soil organic carbon content (%); $Sn1 = 1 - \text{Sand}/100$. Generally, a higher value of the soil erodibility factor K indicates that the soil is poorly resistant to erosion and susceptible to erosion; conversely, the soil is not susceptible to erosion [32–34].

(3) km LS value of slope length factor

The slope length factor, i.e., the topographic factor, determines the state and direction of movement of surface runoff [35]. The greater the slope and the longer the slope length, the greater the potential energy that surface runoff will acquire, and the more intense the erosive effect on soil. In this study, the slope and slope length factors were extracted by the formulas studied by McCool et al. [36] and Liu et al. [37]. The calculation formulae of slope factors are as follows:

$$S = \begin{cases} 10.8 \cdot \sin\theta + 0.03 & \theta < 6 \\ 16.8 \cdot \sin\theta - 0.50 & 5 \le \theta < 14 \\ 21.91 \cdot \sin\theta - 0.90 & \theta < 14 \end{cases} \tag{4}$$

where S is the slope factor (dimensionless), and θ is the slope value (°), which can be extracted from the DEM data.

The formula for calculating the slope length factor is as follows:

$$L = (\lambda/22.13)^{\alpha} \tag{5}$$

$$\lambda = flowacc \times cellsize \tag{6}$$

$$\alpha = \beta/(1+\beta) \tag{7}$$

$$\beta = (\sin\theta/0.089)/\left[3.0 \times (\sin\theta)^{0.8} + 0.56\right] \tag{8}$$

Among them, L is the slope length factor, and its value is the amount of soil erosion produced on the standard slope of 22.13 m. The λ is the slope length, where *flowacc* is the catchment accumulation, *cellsize* is the size of the DEM data grid pixel, and α is the slope length, θ is the slope value, in units of (°); β is the parameter that determines α.

(4) C value of vegetation cover and management factor

Vegetation can protect the surface soil and slow down the rate of soil erosion [38]. NDVI is the most common data to calculate the C value of vegetation cover and management factor [39]. The NDVI number 9 used in this study is derived from the Google Earth Engine cloud platform, and the formula proposed by VanderKnijff et al. [40] is used to calculate the C value of vegetation cover and management factor. The formula is as follows:

$$C = \exp\left[-a \times \frac{NDVI}{b - NDVI}\right] \tag{9}$$

Among them, C is the vegetation cover and management factor (dimensionless); a and b are the parameters that determine the NDVI-C relationship curve. Through VanderKnijff experiments, it is found that the most appropriate values are $a = 2$ and $b = 1$. This method has been studied in China and has achieved good results. According to the Formula (9), if the C value is negative, the assignment is 0 for all negative values; if the C value is greater than 1, the assignment is 1 for all values greater than 1. The higher the C value, the worse the vegetation growth; on the contrast, the lower the C value, the better the vegetation growth.

(5) *p*-value of soil and water conservation measures

The factor of soil and water conservation and measures generally refers to the ratio of the amount of soil loss when certain engineering measures are taken in a certain area to the amount of soil loss without engineering measures under the same conditions. Its value ranges from 0 to 1; 0 means that soil erosion will not occur in this area, and 1 means no soil and water conservation measures have been taken [38,41–43].

3. Results

3.1. Calculation of Each Factor in the RUSLE Model

(1) R-value of precipitation erosivity factor

This method has been applied in the western region, and good results have been obtained [28]. The average precipitation erosivity factors in 1995, 2000, 2005, 2010, 2015, and 2018 in Lanzhou were 110.06, 83.20, 71.09, 46.68, 56.97 and 198.61 [MJ·mm/(km^2·h·a)] respectively. Spatially, the precipitation erosivity factors of the North and South Mountains decreased from southeast to northwest in 1995, 2000, 2005, and 2010. The precipitation erosivity factors of the North and South Mountains decreased from the west to the east in 2015 and 2018. The precipitation erosivity factor of the west was greater than that of the east. The erosivity factor of precipitation in 2018 was significantly higher than in other years, mainly because 2018 was an abnormally rainy year. The precipitation was higher than that in previous years (Figure 2).

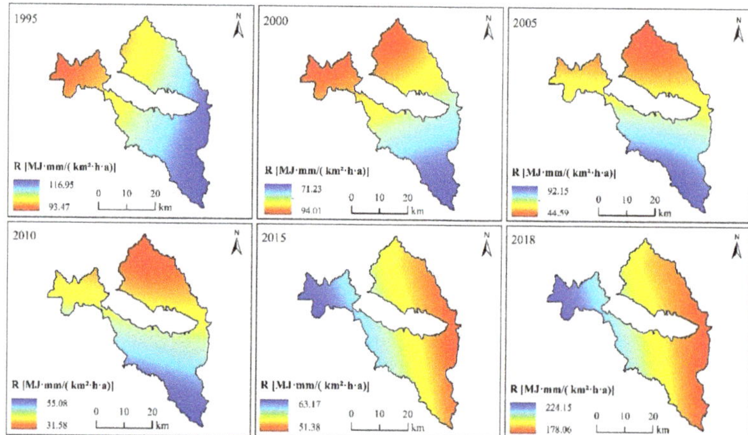

Figure 2. Spatial distribution of rainfall erosivity in the South and North Mountains of Lanzhou in 1995, 2000, 2005, 2010, 2015 and 2018.

(2) K value of soil erodibility factor

According to the data of soil texture and soil organic carbon content of the sampling points, the K value was calculated, and ordinary kriging interpolation was performed in ArcGIS 10.4 software. The spatial distribution of soil erodibility factors in the northern and southern mountains of Lanzhou City was calculated according to Formula (3) (Figure 3). The areas with a soil erodibility factor of 0.054–0.061 t·km^2·h/(km^2·MJ·mm) were mainly distributed in the central and eastern regions, and the areas with ? soil erodibility factor of 0.045–0.053 t·km^2·h/(km^2·MJ·mm) were mainly distributed in the western, northwest and southern regions. The areas with a soil erodibility factor of 0.037–0.044 t·km^2·h/(km^2·MJ·mm) were mainly distributed in parts of the North Mountain. The areas with a soil erodibility factor of 0.018–0.036 t·km^2·h/(km^2·MJ·mm) were mainly distributed in the western part of the North Mountain.

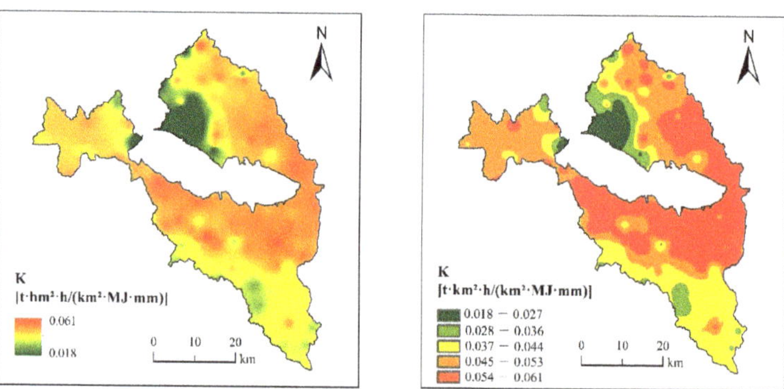

Figure 3. Spatial distribution of soil erodibility factor K in the South and North Mountains of Lanzhou.

(3) LS value of slope length factor

The spatial distribution of the slope factor of the North and South Mountains in Lanzhou (Figure 4) showed that the minimum value of the slope factor was 0, the maximum value was 58.98, the average value was 15.52, the minimum value of the slope factor was 0, the maximum value was 9.99, the average value was 4.76, the minimum value of the slope length factor was 0, the maximum value was 5.92, and the average value was 2.22. The minimum value of the slope length factor was 0, the maximum value was 59.19, and the average value was 12.20. The overall upper slope, slope factor, slope length factor, and slope length factor were zonal distributions, and the slope length factor of the South Mountain was larger than that of the North Mountain.

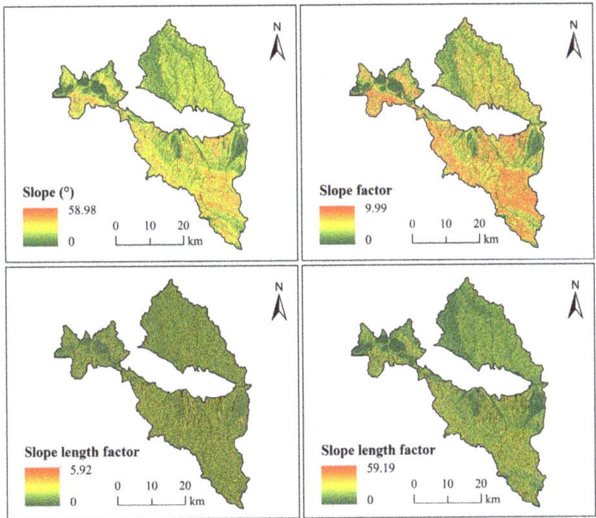

Figure 4. Spatial distribution of the gradient slope and slope length factor in the Northand South Mountains of Lanzhou.

(4) C value of vegetation cover and management factor

The average values of vegetation cover and management factors in the North and South Mountains of Lanzhou in 1995, 2000, 2005, 2010, 2015, and 2018 were 0.34, 0.43, 0.56, 0.50, 0.40 and 0.57, respectively. Overall, the vegetation cover and management factors were the lowest in 1995 and the highest in 2018. The C value of the North Mountain was higher than that of the South Mountain, indicating that the vegetation coverage of the North Mountain was lower than that of the South Mountain (Figure 5).

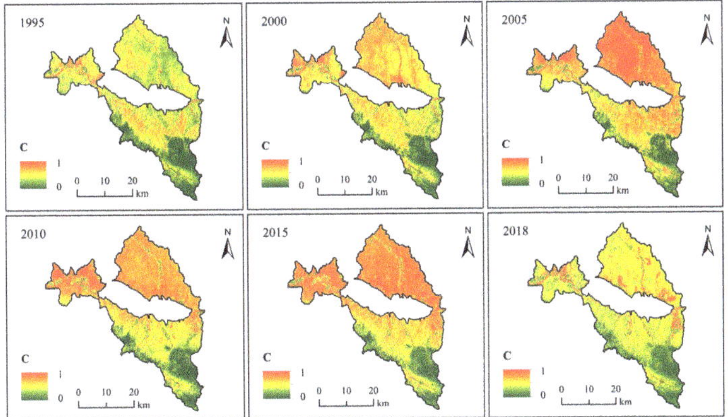

Figure 5. Spatial distribution of vegetation cover and management factors in the South and North Mountains of Lanzhou in 1995, 2000, 2005, 2010, 2015 and 2018.

(5) p-value of soil and water conservation measures

In this study, according to Table 1, the land use data of 1995, 2000, 2005, 2010, 2015, and 2018 were assigned, and the spatial distribution of the p-value of soil and water conservation measure factors with 30 m resolution was obtained (Figure 6). As the land-use change in Lanzhou's North and South Mountains was not obvious, the spatial distribution of soil and water conservation measures in the North and South Mountains was consistent, and the change was not obvious.

Table 1. p values of different land-use types in the South and North Mountains of Lanzhou.

Land-Use Type	Cultivated Land	Forest Land	Grassland	Water Area	Construction Land	Unused Land
p	0.35	1.0	1.0	0.0	0.0	1.0

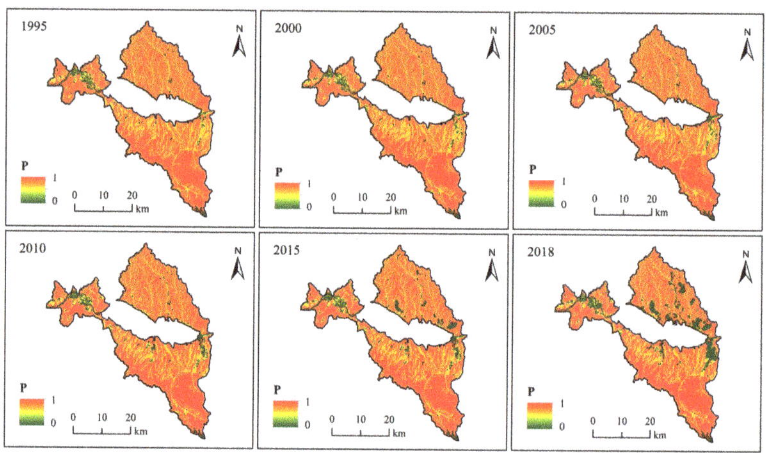

Figure 6. Spatial distribution of soil and water conservation measures factors in the South and North Mountains of Lanzhou in 1995, 2000, 2005, 2010, 2015 and 2018.

3.2. Spatio-Temporal Variation Characteristics of Soil Erosion

According to the soil erosion modulus in different years, the average soil erosion modulus in the North and South Mountains of Lanzhou City generally showed a first fluctuating downward trend and started to increase after 2016, and in 2018 there was an abrupt change and a sudden increase to 25.83 t/(km^2·a). The annual average soil erosion amount was 330.74 × 10^4 t (1995), 323.80 × 10^4 t (2000), 342.09 × 10^4 t (2005), 200.20 × 10^4 t (2010), 314.41 × 10^4 t (2015), and 515.14 × 10^4 t (2018) (Figure 7).

According to the Ministry of Water Resources (SL190-2007) Soil Erosion Classification and Grading Standard [44], the study area is divided into six soil erosion intensity classes according to the soil erosion modulus, namely, slight erosion [5–25 t/(km^2·a)], light erosion [5–25 t/(km^2·a)], moderate erosion [25–50 t/(km^2·a)], strong erosion [50–80 t/(km^2·a)], extremely strong erosion [80–150 t/(km^2·a)] and severe erosion [150 t/(km^2·a)]. The spatial distribution of soil erosion intensity classes for 1995, 2000, 2005, 2010, 2015 and 2018 was obtained for the North and South Mountains of Lanzhou (Figure 8). Soil erosion intensity in the North and South Mountains of Lanzhou in 1995, 2000, 2005, 2010, 2015, and 2018 was mainly slight erosion. The amount of soil erosion in 2018 was significantly higher than that in other years, mainly because 2018 was an unusually wet year, and the precipitation was higher than in previous years. It was mainly distributed in the northwest and southeast of the North and South Mountains. The strong, extremely strong, and severe soil erosion was mainly distributed in the middle of the South Mountain and the middle of the North Mountain (Figure 8).

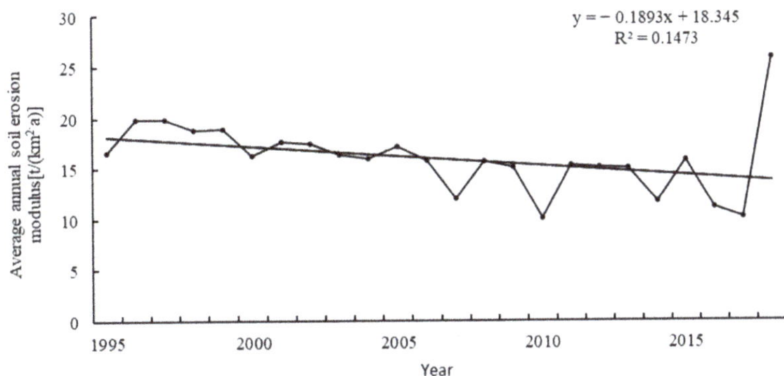

Figure 7. Time change of annual soil erosion modulus in the Northand SouthMountains of Lanzhou from 1995 to 2018.

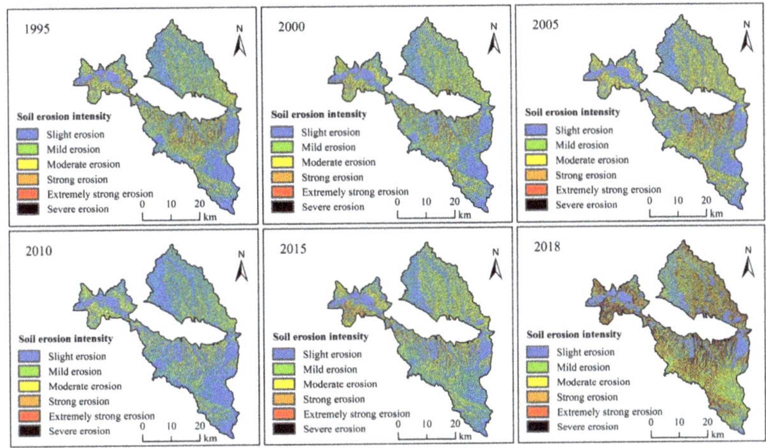

Figure 8. Spatial distribution of soil erosion intensity in the South and North Mountains of Lanzhou in 1995, 2000, 2005, 2010, 2015 and 2018.

3.3. Area Transfer Characteristics of Soil Erosion Intensity

Based on the statistical analysis of the data of different erosion intensity areas, the transfer chord diagram of soil erosion intensity was obtained. From 1995 to 2018, the area of soil slight erosion was the largest, which was 735.92 km^2, followed by mild, moderate, strong, extremely strong, and the area of severe erosion remained the smallest. Among them, slight soil erosion mainly shifted to mild soil erosion. The mild soil erosion transferred to slight soil erosion, and the area was 215.14 km^2; moderate soil erosion mainly shifted to slight and mild soil erosion; strong soil erosion mainly shifted to mild and moderate soil erosion; moderate and strong soil erosion transferred to extremely strong soil erosion; severe soil erosion shifted to strong and extremely strong soil erosion. From 1995 to 2018, the stability rates of slight, mild, moderate, strong, extremely strong and severe soil erosion were 36.91%, 4.64%, 2.74%, 0.79%, 0.77% and 0.08%, respectively (Figure 9).

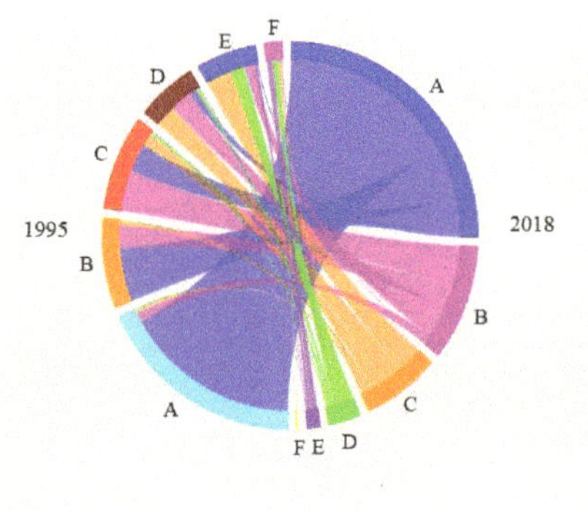

Figure 9. Chordal graph of soil erosion intensity in South and North Mountains of Lanzhou from 1995 to 2018 (Note: A: slight erosion; B: mild erosion; C: moderate erosion; D: strong erosion; E: extremely strong erosion; F: severe erosion).

3.4. Characteristics of Soil Erosion under Different Environmental Factors

3.4.1. Characteristics of Soil Erosion at Different Elevations

The soil erosion modulus in 1995, 2000, 2005, 2010, 2015, and 2018 was superimposed and analyzed according to different elevations, and the average soil erosion modulus at different elevations was obtained (Table 2). It can be seen from the table that the soil erosion modulus first increased and then decreased with the increase in altitude; at the height of 1494–1800 m, the slope length factor is lower, so the soil erosion modulus is lower. At the height of 1800–2100 m, there are more gullies, the slope length factor is high, andcoupled with human interference, so the soil erosion modulus is the largest. With the increase in altitude, the growth of vegetation is better, the vegetation cover and management factors are lower, and coupled with the reduction of human activities, the degree of soil intervention is low. Accordingly, the average soil erosion modulus decreased with altitude.

Table 2. Modulus of soil erosion in different years at different altitudes in the South and North Mountains of Lanzhou (unit: t/(km^2·a)).

Elevation (m)	1995	2000	2005	2010	2015	2018	Average Value
1494–1800	15.05	15.40	15.35	9.92	15.11	31.86	17.11
1800–2100	20.82	19.73	20.44	12.00	19.03	41.21	22.21
2100–2400	18.21	16.58	17.95	9.48	15.07	40.28	19.59
2400–2700	10.36	11.25	14.51	6.98	11.13	32.45	14.45
2700–3000	6.81	7.66	11.03	4.88	8.41	24.02	10.47
3000–3300	2.92	3.95	8.20	3.07	6.52	19.35	7.33
3300–3625	1.26	1.83	3.46	1.56	2.60	6.73	2.91

3.4.2. Characteristics of Soil Erosion under Different Slopes

The soil erosion modulus in 1995, 2000, 2005, 2010, 2015, and 2018 was analyzed according to different slope grades, and the average soil erosion modulus under different slopes was obtained (Table 3). On the whole, the average soil erosion modulus was the highest on the slope of >35°and the lowest on the slope of 0–5°. The 17 modulus of soil erosion increased with the increase in slope. This is mainly because the higher the slope, the greater the slope factor, and the rapid erosion of the runoff velocity caused by surface water is serious.

Table 3. Modulus of soil erosion in different years under different slopes in the South and North Mountains of Lanzhou (unit: t/(km^2·a)).

Slope	1995	2000	2005	2010	2015	2018	Average Value
0–5°	4.19	3.85	3.82	2.20	3.47	8.08	4.27
5–8°	7.09	6.65	6.62	3.82	5.98	13.74	7.32
8–15°	13.55	13.07	13.28	7.66	11.94	27.10	14.43
15–25°	21.40	21.13	22.32	13.08	20.45	46.46	24.14
25–35°	27.44	27.31	30.17	17.95	28.58	65.04	32.75
35–60°	31.46	31.07	35.47	20.90	33.40	76.90	38.20

3.4.3. Characteristics of Soil Erosion under Different Land-Use Types

The average soil erosion modulus under different land-use types was obtained based on the regional statistical analysis of land use classification and soil erosion modulus in 1995, 2000, 2005, 2010, 2015, and 2018 (Table 4). Overall, the average soil erosion modulus of grassland and woodland was larger, which is 24.76 t/(km^2·a) and 23.43 t/(km^2·a), respectively. The average soil erosion modulus of cultivated land was 7.48 t/(km^2·a), and the average soil erosion modulus of water area was 0.27 t/(km^2·a), which is the smallest. Although grassland and woodland are covered by vegetation, grassland and woodland are relatively high above sea level, generally distributed in areas with high mountains and steep slopes, and soil erosion is more serious due to water runoff and gravity.

Table 4. Modulus of soil erosion in different years under different land-use types in the South and North Mountains of Lanzhou (unit: t/(km^2·a)).

Land-Use Type	1995	2000	2005	2010	2015	2018	Average Value
Cultivated land	7.51	6.87	6.83	3.71	5.71	14.26	7.48
Grassland	21.12	22.37	22.42	13.30	21.09	48.25	24.76
Woodland	19.31	20.87	20.97	11.52	19.38	48.53	23.43
Water area	0.32	0.24	0.13	0.11	0.23	0.57	0.27
Construction land	0.47	0.37	0.32	0.19	0.22	0.60	0.36
Unused land	1.18	1.20	1.32	0.58	0.48	2.68	1.24

3.4.4. Characteristics of Soil Erosion under Different Vegetation Coverage

Based on the regional statistical analysis of soil erosion modulus in 1995, 2000, 2005, 2010, 2015, and 2018 according to the different vegetation coverage, the average soil erosion modulus under different land-use types was obtained (Table 5). On the whole, the soil erosion modulus was the largest under low mulch. Except for bare land, the average soil erosion modulus decreased with the increase in vegetation coverage. The rise of vegetation coverage reduced rain water's splashing and running water scouring. The roots of vegetation maintained the soil and played a role in slowing down soil erosion. The average soil erosion modulus of bare land ranked third, mainly because according to the vegetation coverage of less than 10%, and the land use types are construction land and unused land., construction land was generally cement-hardened and difficult to erode; unused land generally had poor soil texture and relatively weak erosion.

Table 5. Modulus of soil erosion in different years under different vegetation coverage in the South and North Mountains of Lanzhou (unit: t/(km^2·a)).

Vegetation Coverage	1995	2000	2005	2010	2015	2018	Average Value
Bare land	21.67	17.29	16.52	11.31	19.66	21.05	15.50
Low vegetation cover	21.21	19.70	21.90	12.92	18.78	43.90	20.06
Medium and low vegetation cover	19.03	19.71	20.82	11.12	15.48	44.21	19.05
Medium vegetation cover	14.42	12.68	16.60	7.58	11.19	38.34	14.97
Medium and high vegetation cover	9.19	7.88	12.35	5.51	9.55	31.47	11.56
High vegetation cover	3.36	3.16	5.16	2.73	7.18	21.64	7.03

3.4.5. Soil Erosion Characteristics under Different Soil Types

In ArcGIS 10.4, the South and North Mountains of Lanzhou City were classified into semi-luvisols, primarosol, pedocal, xerosol, alpine soil and anthrosol according to the definition criteria of the Chinese soil outline [45]. The soil erosion moduli in 1995, 2000, 2005, 2010, 2015 and 2018 were superimposed and analysed according to different soil types to obtain the average soil erosion moduli under different soil types. The average soil erosion modulus was highest for pedocal and lowest for alpine soil (Table 6).

Table 6. Modulus of soil erosion in different years at different soil type in the South and North Mountains of Lanzhou City (unit: t/(km^2·a)).

Soil Type	1995	2000	2005	2010	2015	2018	Average
semi-luvisols	8.47	9.33	12.77	5.79	9.89	29.00	12.54
primarosol	17.94	17.93	18.43	11.38	17.50	38.31	20.25
pedocal	21.63	17.60	21.50	10.75	15.86	41.18	21.42
xerosol	14.10	15.68	14.60	8.76	15.31	31.77	16.70
alpine soil	4.40	6.43	10.12	4.59	6.58	17.83	8.32
anthrosol	15.21	12.35	11.18	8.46	12.67	27.56	14.57

3.4.6. Comparison of Soil Erosion Inside and Outside the Environmental Greening Project

According to the statistics of soil erosion modulus in 1995, 2000, 2005, 2010, 2015 and 2018, the average soil erosion modulus were inside and outside the environmental greening project (Table 7). The average soil erosion modulus inside and outside the environmental greening project was 21.27 t/(km^2 a) and 23.56 t/(km^2 a), respectively, and the average soil erosion modulus outside the environmental greening project was larger than that inside the environmental greening project. The area of slight soil erosion inside and outside the greening area was the largest, and the area of severe soil erosion was the smallest, which was 7.72 km^2 and 4.11 km^2, respectively, accounting for 0.39% and 0.21% of the total area. Within and outside the greening range, the area occupied by soil erosion intensity from large to small was slight, mild, moderate, strong, extremely strong and severe, respectively (Figure 10).

Table 7. Modulus of soil erosion in different years inside and outside the environmental greening project in the South and North Mountains of Lanzhou City (unit: t/(km^2·a)).

	1995	2000	2005	2010	2015	2018	
inside	18.81	18.55	19.74	11.98	18.36	40.19	21.27
outside	22.28	21.79	22.19	15.29	20.57	39.23	23.56

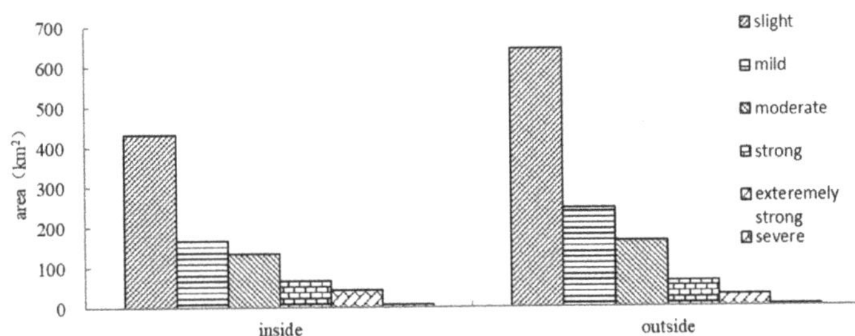

Figure 10. Average area of soil erosion intensity inside and outside the environmental greening project of South and North Mountains in Lanzhou City (unit: km^2).

4. Discussion

From 1995 to 2018, the average soil erosion modulus in the North and South Mountains of Lanzhou showed a fluctuating downward trend. There was an abnormality in 2018, mainly due to precipitation. The natural precipitation in the North and South Mountains of Lanzhou was less and the seasonal distribution is uneven. The precipitation was mostly concentrated in July, August, and September, and its rainfall accounts for 60–70% of the annual rainfall. However, 2018 was a year of abnormal increase in precipitation. Although the average precipitation was 441.6 mm, which is less than that in other areas, the rainfall erosivity was the highest due to the low vegetation coverage and the collapsibility of loess. Torrential rain caused flash floods, landslides, and mudslides, resulting in the largest in soil erosion modulus in 2018.

Due to climate, natural disasters, and man-made deforestation, soil erosion and desertification in the study area are becoming more and more serious. Lanzhou's urban construction is closely related to the ecological construction of the North and South Mountains. In order to improve its ecological environment, Lanzhou started planting plantation greening projects in 1999. Therefore, most of the existing forests in Lanzhou are plantations, and a preliminary plantation ecosystem has been formed. At present, the method of increasing vegetation coverage and reducing the bare area of the ground has achieved initial results, and the soil erosion modulus within the greening project is lower than that outside the greening project.

Soil erosion moduli can be determined in a variety of ways, such as using measured runoff sediment information, simulated rainfall, field surveys, radioisotopes, and mathematical models. In this paper, the soil erosion modulus was calculated using the Revised Universal Soil Loss Equation (RUSLE). Although the limitations of the application of RUSLE in the North and South Mountions of Lanzhou City were corrected to the greatest extent possible, the RUSLE can only calculate the hydraulic erosion modulus, and parts of the North and South Mountains are in a wind erosion zone with a large area of desertification and dust storms occurring every spring [46]. As there is still a lack of effective calculation models for wind and water erosion composite soil erosion modulus, the changes in wind erosion modulus are not considered in this paper, resulting in a small calculation of soil erosion modulus in windy and sandy areas. Soil erosion is closely related to land desertification, and effective management of soil erosion can help to slow down the development of land desertification in the study area [47,48].

Countermeasures to prevent soil erosion in the North and South Mountains are the following:

(1) Reasonably control the development of land resources in the North and South Mountains and improve the compensation mechanism for land resource protection.
(2) Strengthen the construction of scientific field observation and research stations for the ecological environment in the North and South Mountains of Lanzhou.

(3) Increase investment in environmental greening project in the North and South Mountains and consolidate the role of the ecological security barrier in the North and South Mountains.

5. Conclusions

(1) The average precipitation erosivity factors of the North and South Mountains of Lanzhou in 1995, 2000, 2005, 2010, 2015, and 2018 were 110.06, 83.20, 71.09, 46.68, 56.97, and 198.61 MJ·mm/(km^2·h), respectively. Spatially, precipitation erosivity of the North and South Mountains decreased from southeast to northwest in 1995, 2000, 2005, and 2010, and decreased from west to east in 2015 and 2018.
(2) The average soil erosion modulus of the North and South Mountains of Lanzhou fluctuated and decreased from 1995 to 2018. The intensity of soil erosion in 1995, 2000, 2005, 2010, 2015, and 2018 was mainly slight erosion, distributed primarily in the northwest and southeast of the North and South Mountains. Strong, extremely strong, and severe soil erosion was distributed primarily in the middle of the South Mountain, and a small amount in the middle of the North Mountain.
(3) The soil erosion modulus of the North and South Mountains of Lanzhou increased at first and then decreased with the increase in height, increased with the increase in slope, and decreased with the increase in vegetation coverage. Among the land use types, the average soil erosion modulus of grassland and woodland was larger, and that of the water area was the lowest. The soil erosion moduli were greatest in the pedocal of the North and South Mountains, and the least in the alpine soil. The greening project effectively prevented soil erosion and initially formed an artificial ecosystem.

Author Contributions: Conceptualization and Methodology: H.Z.; Writing–original draft, review, and editing: J.L.; Formal analysis: H.W.; Data Curation: C.X.; Visualization: Y.Y. All authors have read and agreed to the published version of the manuscript.

Funding: The research was funded by the Innovation and Entrepreneurship Talent Project of Lanzhou (2019-RC-105), and the National Natural Science Foundation of China (41461011).

Institutional Review Board Statement: Not applicable.

Informed Consent Statement: The study did not involve humans.

Data Availability Statement: Not applicable.

Acknowledgments: This research received help from Chen Lei, Zhang Yuhong, An Huimin, Song Jinyue, Li Ming, and Han Wuhong from field design, sampling, and laboratory data measurement.

Conflicts of Interest: The funders had no role in the design of the study; in the collection, analysis, or interpretation of data; in the writing of the manuscript, or in the decision to publish the results.

References

1. Oldeman, L.R. *Global Extent of Soil Degradation. Bi-Annual Report 1991–1992/ISRIC*; ISRIC: Wageningen, The Netherlands, 1992; pp. 19–36.
2. Liu, H. Types and Characteristics of Land Degradation and Countermeasures in China. *Nat. Resour.* **1995**, *4*, 26–32.
3. Bridges, E.M.; Oldeman, L.R. Global assessment of human-induced soil degradation. *Arid. Soil Res. Rehabil.* **1999**, *13*, 319–325. [CrossRef]
4. Qiao, Y.L.; Qiao, Y. Fast soil erosion investigation and dynamic analysis in the loess plateau of China by using information composite technique. *Adv. Space Res.* **2002**, *29*, 85–88.
5. Vrieling, A.; Sterk, G.; Vigiak, O. Spatial evaluation of soil erosion risk in the West Usambara Mountains, Tanzania. *Land Degrad. Dev.* **2006**, *17*, 301–319. [CrossRef]
6. Zhang, X.W.; Zhou, Y.M.; Li, X.S.; Yuan, C.; Yan, N.N.; Wu, B.F. A Review of Remote Sensing Application in Soil Erosion Assessment. *Chin. J. Soil Sci.* **2010**, *41*, 1010–1017.
7. Zou, Y.; Zhang, J.; Li, G.; Shen, Z.; Lu, F.; Zhou, X. The study on dynamic characteristics of soil erosion in Yuyao City of Zhejiang Province. *IOP Conf. Ser. Earth Environ. Sci.* **2017**, *81*, 012110. [CrossRef]
8. Youcef, S.; Elhadj, M.; Belkacem, M.; Benoit, L.; Christophe, V.; Khodir, M. Mapping surface water erosion potential in the Soummam watershed in Northeast Algeria with RUSLE model. *J. Mt. Sci.* **2019**, *16*, 1606–1615.

9. Liao, H. Soil Erosion has Become the Number one Environmental Problem. In Proceedings of the Tenth World Day to Combat Desertification and Drought, Bonn, Germany, 17 June 2004; p. 3.
10. Ni, J.R.; Li, X.X.; Borthwick, A.G.L. Soil erosion assessment based on minimum polygons in the Yellow River basin, China. *Geomorphology* **2008**, *93*, 233–252. [CrossRef]
11. Zheng, F.; Wang, Z.; Yang, Q. The Retrospection and Prospect on Soil Erosion Research in China. *J. Nat.* **2008**, *30*, 12–16.
12. Ministry of Water Resources of the People's Republic of China. 2018 Soil and Water Conservation Bulletin [EB/OL]. 2018. Available online: http://www.mwr.gov.cn/ (accessed on 4 September 2019).
13. Qu, J.J.; Ma, L.P.; Liu, C. Present Situation, Cause and Control Way of Sandy Desertification in Gansu Province. *J. Desert Res.* **2002**, *22*, 520.
14. Wang, S.G.; Yang, M.; QI, B.; Xin, C.L.; Yang, M.F. Influence of Sand-dust Storms Occuring over the Gansu Hexi District on the Air Pollution in Lanzhou City. *J. Desert Res.* **1999**, *19*, 354.
15. Tao, Y.; Yang, D.R.; Lan, L.; Wang, H.X.; Wang, S.G. Relationship between air pollution and respiratory disease hospitalization in Lanzhou. *China Environ. Sci.* **2013**, *33*, 175–180.
16. Zhao, K.C.; Qu, L.B. Strategy of Vegetation Restoration in Lanzhou North-South Mountains. *J. Desert Res.* **2006**, *3*, 493–497.
17. Wu, Q.L. The Applicability Analysis of Massive Artificial Forestation in Lanzhou South-north Hills. *Res. Soil Water Conserv.* **2003**, *10*, 134–136.
18. Li, J. *Research on the Dynamic Changes of Vegetation Coverage in the North and South Mountains of Lanzhou Based on GIS and RS*; Northwest Normal University: Lanzhou, China, 2009.
19. Renard, K.G.; Ferreira, V.A. RUSLE model description and database sensitivity. *J. Environ. Qual.* **1993**, *22*, 458–466. [CrossRef]
20. Da Cunha, E.R.; Bacani, V.M.; Panachuki, E. Modeling soil erosion using RUSLE and GIS in a watershed occupied by rural settlement in the Brazilian Cerrado. *Nat. Hazards* **2017**, *85*, 851–868. [CrossRef]
21. Wang, T. *Effects Dynamics Change of LUCC on Soil Erosion Intensity in Lushan-Huangling Area*; Northwestern University: Evanston, IL, USA, 2018.
22. Cao, X.Y. *Medium and Long Term Spatio-Temporal Dynamics Analysis of Soil Erosion on Small Watershed*; Northwestern University: Evanston, IL, USA, 2018.
23. Meyfroidt, P. Trade-offs between environment and livelihoods: Bridging the global land use and food security discussions. *Glob. Food Secur.* **2018**, *16*, 9–16. [CrossRef]
24. Renard, K.G.; Foster, G.R.; Weesies, G.A.; Porter, J.P. RUSLE: Revised universal soil loss equation. *J. Soil Water Conserv.* **1991**, *46*, 30–33.
25. Chen, S.X.; Yang, X.H.; Xiao, L.L.; Cai, H.Y. Study of Soil Erosion in the Southern Hillside Area of China Based on RUSLE Model. *Resour. Sci.* **2014**, *36*, 1288–1297.
26. Kayet, N.; Pathak, K.; Chakrabarty, A.; Sahoo, S. Evaluation of soil loss estimation using the RUSLE model and SCS-CN method in hillslope mining areas. *Int. Soil Water Conserv. Res.* **2018**, *6*, 31–42. [CrossRef]
27. Wischmeier, W.H.; Smith, D.D. *Predicting Rainfall Erosion Losses—A Guide to Conservation Planning*; Science and Education Administration, U.S. Dept. of Agriculture: Washington, DC, USA, 1978; p. 537.
28. Gao, H.D.; Li, Z.B.; Li, P.; Jia, L.L.; Xu, G.C.; Ren, Z.P.; Pang, G.W.; Zhao, B.H. The capacity of soil loss control in the Loess Plateau based on soil erosion control degree. *Acta Geogr. Sin.* **2015**, *70*, 1503–1515. [CrossRef]
29. Men, M.X.; Zhao, T.K.; Peng, Z.P.; Yu, Z.R. Study on Soil Erodibility Based on the Soil Particle-size Distribution in Hebei Province. *China Agric. Sci.* **2004**, *37*, 1647–1653.
30. Jiang, X.S.; Pan, J.J.; Yang, L.Z.; Bu, Z.H. Methods of Calculation and Mapping Soil Erodibility K-a case study of FangBain Watershed of Nanjing. *Soils* **2004**, *1*, 177–180.
31. Williams, J.R.; Renard, K.G.; Dyke, P.T. EPIC: A new method for assessing erosion's effect on soil productivity. *J. Soil Water Conserv.* **1983**, *38*, 381–383.
32. Fu, S.; Zha, X. Study on Predicting Soil Erosion in Dongzhen Watershed Based on GIS and USLE. *Geo Inf. Sci.* **2008**, *10*, 390–395.
33. Lu, J.Z.; Chen, X.L.; Li, H.; Xiao, J.J.; Yin, J.M. Soil erosion changes based on GIS/RS and USLE in Poyang Lake basin. *Trans. Chin. Soc. Agric. Eng.* **2011**, *27*, 337–344.
34. Zhu, C.G.; Li, W.H.; Li, D.L.; Liu, Z.; Fu, L. Feature analysis of soil physical properties and erodibility of the Yili Valley. *Resour. Sci.* **2016**, *38*, 1212–1221.
35. Zingg, A.W. Degree and length of land slope as it affects soil loss in runoff. *Agric. Eng.* **1940**, *21*, 59–64.
36. McCool, D.K.; Foster, G.R.; Mutchler, C.K.; Meyer, L.D. Revised Slope Length Factor for the Universal Soil Loss Equation. *Trans. Asae* **1989**, *30*, 1387–1396. [CrossRef]
37. Liu, B.Y.; Zhang, K.L.; Yun, X. An Empirical Soil Loss Equation. In Proceedings of the 12th International Soil Conservation Organization Conference, Beijing, China, 26–31 May 2002.
38. Wang, M.; Liu, Y.; Song, C.; Li, C.L.; Xiao, W.F. Evaluation Soil Erosion Based on RUSLE Model in Three Gorges Reservoir Area During 2000–2010. *Bull. Soil Water Conserv.* **2018**, *38*, 12–17.
39. Zha, L.S.; Deng, G.H.; Gu, J.C. Dynamic changes of soil erosion in the Chaohu Watershed from 1992 to 2013. *Acta Geogr. Sin.* **2015**, *70*, 1708–1719.
40. Van der Knijff, J.M.; Jones, R.J.A.; Montanarella, L. *Soil Erosion Risk: Assessment in Europe*; European Soil Bureau (ESB): Ispra, Italy, 2000.

41. Lu, C.H.; Dai, F.Q.; Liu, G.B. Dynamic Changes of Soil Erosion in the Three Gorges Reservoir Area from 2000 to 2012-A Case Study of Wanzhou District of Chongqing City. *Bull. Soil Water Conserv.* **2017**, *37*, 1–8.
42. Cheng, X.F.; Yu, F. Spatial distribution of soil erosion and its relationship with environmental factors in Anhui Province. *Geogr. Res.* **2010**, *29*, 1461–1470.
43. Yi, K.; Wang, S.Y.; Wang, X.; Yao, H.L. Analysis of spatial and temporal differentiation characteristics of soil erosion based on RUSLE model—A case study of Chaoyang City, Liaoning Province. *Geogr. Sci.* **2015**, *35*, 365–372. [CrossRef]
44. *SL190-2007*; Ministry of Water Resources of the People's Republic of China. Soil Erosion Classification and Grading Standard. China Water Conservancy and Hydropower Press: Beijing, China, 2007.
45. Zhang, G.L.; Wang, Q.B.; Zhang, F.G.; Wu, K.; Cai, C.; Zhang, M.; Li, D.; Zhao, Y.; Yang, J. Criteria for establishment of soil family and soil series in Chinese soil taxonomy. *J. Soil Sci.* **2013**, *50*, 826–834.
46. Zong, Y.W. Analysis of sand and dust storm weather characteristics in Lanzhou City. *Mod. Agric. Sci. Technol.* **2019**, *19*, 184–185.
47. Wang, T.; Zhu, Z.D. Study on Sandy Desertification in China—1. *Definitoin of Sandy Desertification and its Connotation. J. Desert Res.* **2003**, *23*, 209.
48. Zhu, Z.D. Current status and prospects of desertification research in China. *J. Geogr.* **1994**, *1*, 650–659.

MDPI

Opinion

An Artificial Oasis in a Deadly Desert: Practices and Enlightenments

Ying Zhao [1,2,*], Jie Xue [2,3,4], Nan Wu [1,*] and Robert Lee Hill [5]

1 Yantai Key Laboratory of Coastal Ecohydrological Processes and Environmental Security, College of Resources and Environmental Engineering, Ludong University, Yantai 264025, China
2 National Engineering Technology Research Center for Desert-Oasis Ecological Construction, Xinjiang Institute of Ecology and Geography, CAS, Urumqi 830011, China; xuejie11@ms.xjb.ac.cn
3 Cele National Station of Observation and Research for Desert-Grassland Ecosystems, Cele 848300, China
4 University of Chinese Academy of Sciences, Beijing 100049, China
5 Department of Environmental Science and Technology, University of Maryland, College Park, MD 20742, USA; rlh@umd.edu
* Correspondence: yzhaosoils@gmail.com (Y.Z.); wunan@ldu.edu.cn (N.W.)

Abstract: Building highway and its biological protection system in a drought-affected shifting-sand desert is a great challenge. This challenge was completed by the construction of the Taklimakan Desert Highway Shelterbelt (TDHS)—the longest of its kind in the world (436 km). The TDHS can serve as a model for highway construction and desertification control using eco-friendly and cost-effective approaches in other desert regions. Notably, we proved that local saline groundwater irrigation offers potential advantages and opportunities for the growth of halophytes and sandy soil development in hyper-arid desert environments. Here, we systematically (1) summarize the project, its results, and vital technical issues of saline water irrigation; (2) address soil hydrological processes that play a crucial role in maintaining those systems; and (3) highlight useful insights for soil development, plant survival, and soil–plant–water–biota synergy mechanisms. Indeed, the TDHS project has provided a proof of concept for restoration and desert greening initiatives.

Keywords: soil hydrological processes; saline water irrigation; desertification control; shelterbelt

Citation: Zhao, Y.; Xue, J.; Wu, N.; Hill, R.L. An Artificial Oasis in a Deadly Desert: Practices and Enlightenments. *Water* **2022**, *14*, 2237. https://doi.org/10.3390/w14142237

Academic Editors: Jan Wesseling and Guido D'Urso

Received: 1 June 2022
Accepted: 14 July 2022
Published: 16 July 2022

1. Introduction

Over the last 20 years, a great "green wall" of 436 km has been gradually erected in the center of the Taklimakan Desert, and has helped transform the desert into an oasis [1]. People could not travel very far into the desert in previous decades and centuries because of its hyper-arid environment, known as the "sea of death" in China. It seems that human ingenuity and science can overcome tremendous environmental barriers of fiercely dry and hot conditions to develop new, green ecosystems that provide adaptation services such as protection of infrastructure. The possibility of vegetating large sand-dune deserts as a way to transform their climate to more habitable states—providing global benefits such as carbon sequestration—is increasingly discussed and explored through paleo-evidence and modeling [2]. Here, we report this singular eco-engineering experience (the Taklimakan Desert Highway Shelterbelt (TDHS)), offering proof of how human-aided vegetation establishment on hyper-arid mobile dunes can start and persist at massive scales.

In some contexts, this project may be a successful artifact of humanity's adaption to nature. In China, it has been well known for a thousand years that the Dujiangyan Irrigation System represents a remarkable milestone in human design, adapting to nature. The Chinese government has occasionally examined seawater transport from the eastern coast for use in the Taklimakan Desert. While long-distance water transport might seem impractical, the South-to-North Water Diversion Project to transport water from southern to northern China has caused renewed consideration of such an engineering feat. In Jordan, there is currently an ambitious project to pipe saltwater from the Red Sea to the arid coastal city of Aqaba. The goal is to turn the region into an oasis using seawater that has been desalinated by combining seawater greenhouses and concentrated solar power [3].

In contrast, the Taklimakan "experiment" shows concrete and large-scale evidence that the seed of such transformation is technically feasible by employing local saline groundwater. The portfolio-wide planning of the TDHS may directly feed into the realization of the United Nations' 2030 Agenda for Sustainable Development, due to its potential for carbon sequestration and eco-restoration [4]. These steps would enable the evaluation of outcomes in different social–ecological systems (SESs), e.g., the conditions of ecosystem services, choices of livelihood strategies, and cultural values [5]. Modern culture has developed the techniques to meet the challenges of coordinating or overcoming extreme natural conditions.

The desert is a unique ecosystem. Even though we have modern technology to available re-cultivate our planet, the question remains as to whether this is a wise choice, and why we would celebrate transforming this—one of the last wild places on Earth—into a more artificial territory. The practicality of creating oases as geographic features within desert areas differs from technically feasible. The significance of ecosystem rehabilitation and management requires supportive and regulatory ecosystem services, ecosystem sustainability, and land–atmosphere feedbacks. Most eco-restoration efforts are in temperate regions where the water demands of planted forests can be of concern because the planted trees can exhibit higher evapotranspiration rates than pre-afforestation farmland and grassland [6]. Historically, the unreasonable anthropogenic activity associated with natural shifts has significantly impacted our habitat environment, and has even led to the fall of more than one civilization. For instance, limited understanding of land and water relations left the Tigris and Euphrates valleys saline and barren. Nevertheless, there is no doubt that the existence of vegetation in sandy deserts is influential. Understanding the two-way coupling between vegetation dynamics and the water cycle is critical for the sustainable management of water-limited landscapes [2].

Our study shows the importance of controlling and reclaiming areas overtaken by desert. The most successful context-specific technologies for desert highway and protection systems illustrate how China has tackled those challenges, and provide invaluable guidance for other nations embarking on a similar journey. Given the TDHS having attracted widespread attention concerning saline water utilization and the use of breakthrough technological applications within areas in which standard ecological engineering practices may not apply, this perspective systematically reviews the effects of the TDHS on local saline groundwater utilization, and how it has evolved over time, highlighting valuable insights for fundamental soil hydrological processes associated with soil development and plant survival. We also propose some core principles for successfully implementing such large-scale nature-based solutions with consideration of technology transfer.

2. Overview of TDHS

The TDHS was constructed in 2003, as the initial mechanical checkerboard failed to prevent shifting sands from blocking the highway built in 1997 to access the oil fields [7,8] (Figure 1). However, before the construction of the TDHS, there was no extensive ecological engineering on projects in shifting desert areas. It is essential to select and cultivate appropriate plant species that are drought- and salt-tolerant for wind erosion prevention, and to develop an integrated, innovative irrigation system for afforestation. In 1999, a pilot test was started to assess the feasibility of highway shelterbelt construction. A botanical garden was built within the center of the oasis for trait-based plant species selection and public viewing of the pilot test (Figure 1). By introducing plants that have been pilot-tested in the central oasis, the mobile dunes along the TDHS have been successfully stabilized over the years (Figure 1). At present, a 72–78-meter-wide tree belt has been built along the 436 km of the highway, with a green area of 3128 ha. Despite groundwater with a high saline content, the greenbelt has been most successful, with a 90% plant survival rate [1].

Furthermore, its establishment has improved the biological diversity and soil fauna activity, e.g., 13 herb species have been found since the construction of the TDHS [9].

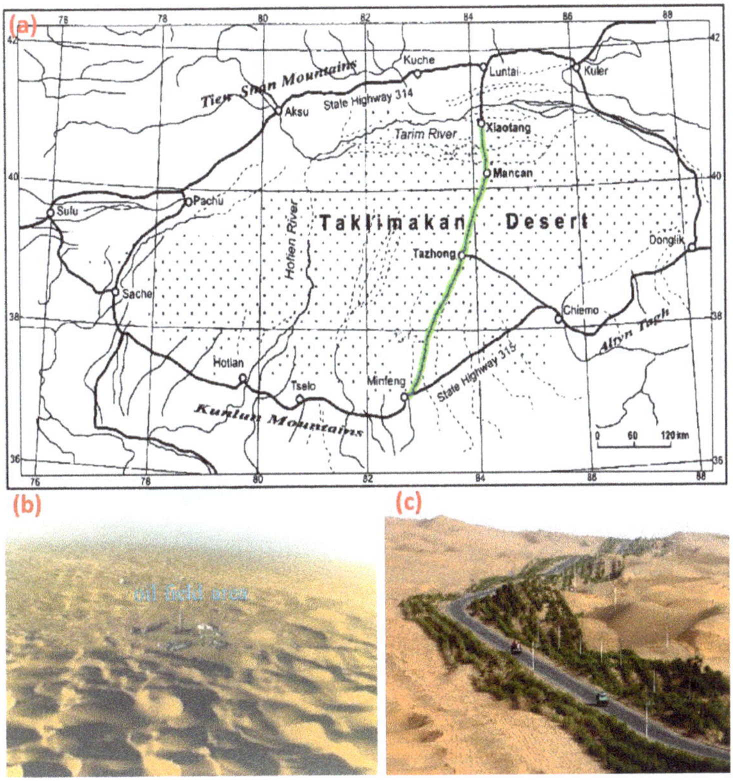

Figure 1. The Taklimakan Desert Highway Shelterbelt across a shifting-sand desert: The highway links Luntai and Minfeng on the northern and southern edges of the Tarim Basin (562 km). The greenbelts on both sides of the highway represent the artificial vegetation establishment (**a**), the oil field area—more than 98% of areas were covered with shifting dunes before the shelterbelt's construction along the Taklimakan Desert Highway (**b**), and the greenbelt today (**c**).

The TDHS's success relied on modern engineering, irrigation, and soil management technologies. First, and most importantly, local saline groundwater irrigation offers advantages and opportunities for plant growth and soil evolution in the mobile sand environment. Large-scale afforestation was possible using specially designed double-branch-pipe drip-irrigation systems to overcome the limitation of regional water scarcity, triggering a rapid evolution of sandy soils that favored further long-term vegetation rehabilitation. We consider saline irrigation to be a feasible way to initialize the sustainability of artificial shelterbelts in similar habitats. What was originally a tree-planting initiative for road construction has evolved into China's response to climate change and a broader development programming tool. Since it was created in a dry desert and nourished with local groundwater, the TDHS project has been highly appraised by many national and international environmental protection organizations—for instance, receiving the highest award of the Chinese Environmentally Friendly Project in 2008. Its success was attributed to modern technologies in engineering design, irrigation utilization, and unique/localized management practices.

3. Soil Hydrological Processes under Irrigation in the Desert

3.1. Saline Water Irrigation for Shelterbelt Engineering in Desert

The near-impossible challenge that this project overcame was the issue of ensuring water sources. The TDHS is irrigated by local groundwater resources, although the water in question is saline. Our observation showed that irrigation with saline groundwater with a salinity concentration of <15.5 g L^{-1} has allowed the plants to thrive with the trend of soil salinity [10]. Due to salt accumulation outside of the root zone, saline groundwater irrigation has not harmed the selected plants' healthy growth. Moreover, the saline water irrigation seems to have provided nutrients to the soil and, thus, stimulated soil development [8]. Consequently, the ecological restoration has increased the silt and clay content, thereby enhancing the water-holding capacity, which is one of the vital prerequisites for long-lasting plant colonization within desert ecosystems [11]. These synergistic mechanisms might be particularly crucial with respect to the initial sandy soil's evolution and the survival of sand-binding vegetation.

Drought, drifting sand blockages, and salt accumulation are three factors that may threaten the sustainability of the artificial desert oasis. In general, precipitated water will all evaporate in the desert air, and the water sufficiency of oasis plants is questionable. Although a water-saving drip-irrigation system was used for our oasis, the soil water content was usually less than 5%, due to the low water-holding capacity of the sandy soil and high evaporative demands. More frequent irrigation schedules (once per 1.5 weeks) have proven better for dealing with drought [12] (Figure 2). Our studies have demonstrated that saline irrigation could significantly influence the dynamics and distribution of soil moisture and salt [13]. Given the buildup of soil salinity through irrigation with saline water, sustainable irrigation strategies are critical to reducing ecological risks [14]. In our case, where an aquifer might not have enough pressure to break through to the surface, a well was bored for pumping water to the surface to irrigate the newly planted trees. This kind of action can ensure the survival of an oasis, and can help jumpstart (i.e., catastrophic shifts) a natural spring that continues to feed an oasis long into the future.

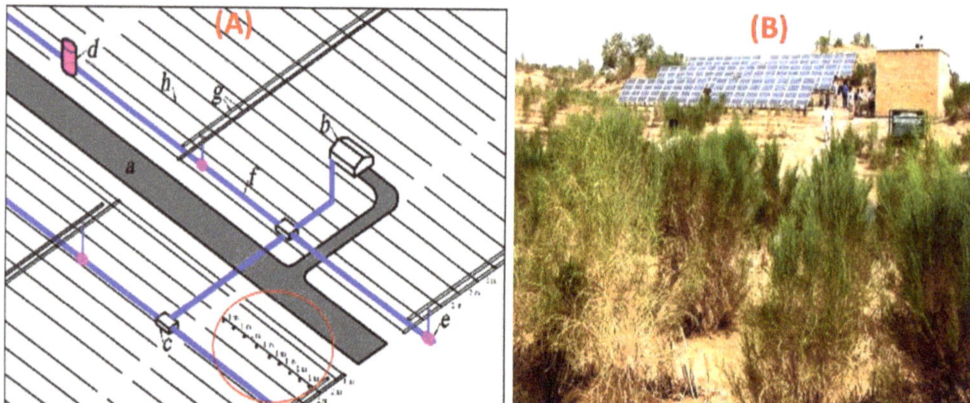

Figure 2. The drip-irrigation system for the Taklimakan Desert Highway Shelterbelt (**A**), which is powered by solar panels in some sections (**B**). (a) Desert highway; (b) well-house; (c) control valve well; (d) connecting well; (e) water valve; (f) main pipe; (g) double-branch pipe; (h) capillary tube. Figure 2 is sourced and updated from [13].

Moreover, a complicated but plausible explanation may be that the decreased Na^+ content was replaced by Ca^{2+} within the root-zone soil, as follows (Figure 3): (i) halophyte root respiration and decomposition of organic matter produced a high partial pressure of O_2 and CO_2 [15]; (ii) the reaction of CO_2 with H_2O (irrigation water and soil moisture) produced H_2CO_3 [16]; (iii) the dissociation of H_2CO_3 and/or N_2-fixing of halophyte roots released H^+; (iv) the reaction of H^+ with $CaCO_3$ resulted in the Ca^{2+}; and (v) the Ca^{2+} facilitated the removal of Na^+. Consequently, the Na^+ may eventually be removed through uptake by halophytes and/or leaching with irrigation water. Although this somewhat complicated process needs additional study, such a process would help explain the decreased soil salinity.

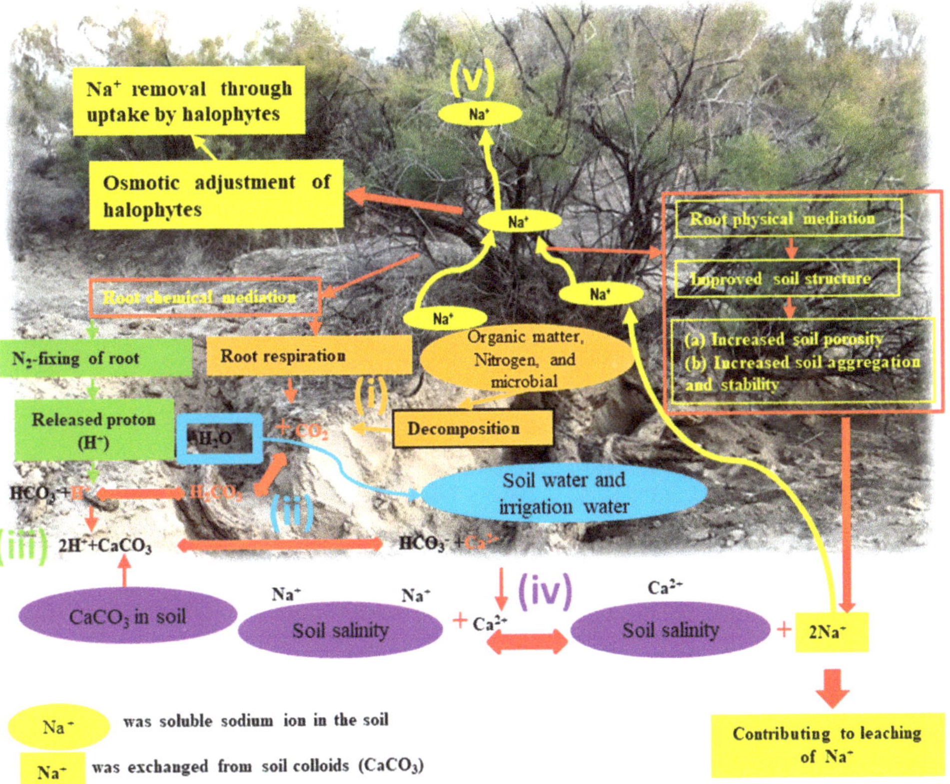

Figure 3. Concept of a suitable mechanism of halophytes' rhizospheres on salt-affected sandy soil under saline water irrigation (an evacuated soil profile of *Tamarix taklamakanensis* is in the background). The figure is referenced and updated from [16].

3.2. Dominant Soil Hydrological Processes under Irrigation Conditions

The local saline groundwater with a salt content of 5–30 g L^{-1} was used to irrigate the TDHS. During drip-irrigation, the salts were transported upward with water for evaporation, and with three dimensions for equilibration of soil water potential differences. The upward movement may lead to salt accumulation at the topsoil. However, this surface salt-accumulation process would weaken, because salty surface crusts may block evaporation [13]. Thus, surface salinization may eventually stop at some point. With a periodically large amount of drip-irrigation, downward water flow would make salts move deeper and prevent their accumulation at a certain depth. Eventually, the salt and water would reach a relatively shallow groundwater level (e.g., 3–5 m) and finish the local water cycle [8]. Recent research also suggests that combining the natural freshwater and desalinated seawater is an excellent strategy to respond to the high water demands for crops [17], along with water pretreatment to reduce salinity as the magnetized, oxygenized, or electrical water approaches.

It should be noted that long-term saline water irrigation frequently results in soil salt accumulation that is detrimental to plant growth. However, our field observations tell a different story. Sandy soil has a low absorption rate, and the saline solutes travel far below the tree roots. The soil EC increased by about 8 mS cm^{-1} within the 0–10 cm soil layer, where no lateral or feeder roots existed, and was less than 1.0 mS cm^{-1} within the 40–60 cm soil layer, where plant roots proliferated. Due to those differences between the salty soil layer and the root zone, our monitoring results showed that saline water irrigation did not harm the plants that had adapted to the local environment.

4. Implications of the TDHS Project

4.1. Underlying Mechanisms to Maintain an Artificial Ecosystem in the Desert

It should be mentioned that the soil ecology has shifted from barren land to vegetated area, which is a significant contributor to sustainability. For >10 years since the TDHS began construction, we have found that the soil structure has progressively improved, as indicated by increased soil aggregate stability and the increased fractal dimensions associated with the increased total N and organic matter contents [18]. The soil is a living miniature ecosystem, exhibiting structured soil processes and functions based on more than sandy particle or soil grain characterization [19]. Yi et al. [20] reported that a rheological state—the so-called omnidirectional integrative constraint "field"—may have formed after adding soil amendment to the sand, providing a prerequisite for soil to become a habitat for plants. However, it has not been determined whether this attribute of binding force exists as a real, measurable variable. Our research found that these amendments improved the shear strength of aeolian soil by enhancing soil cohesion and stabilizing the aeolian soil [21]. Nevertheless, the sand fixation strategy is vital to assist in the governance of loose sand. Additional amendments include reintroducing regulatory species, stimulating soil development with fungi and microorganisms, and re-establishing mutual relationships between species (e.g., vaccination with mycorrhiza) [22].

It could be fascinating for a newly reclaimed ecosystem to apply a structural equation model to understand the relationships between structure, process, and outcome, and to examine the ecosystem services. This is necessary to understand salt-retention mechanisms and soil–plant–water–biota processes further. However, the manifestation of this vast artificial ecosystem remains vague with regard to resource availability and utilization, and to the appropriate management practices for such an enormous heterogeneous area. As a result, the implications of China's integrated land system sustainability remain under-recognized. To identify catastrophic regime shift behavior that results from changes in dryland SESs' structure, functions, and their interactions—and, hence, to determine the tipping point for maintaining ecological sustainability—it is urgently required to unravel the underlying mechanisms and stability from holistic and context-specific perspectives [5].

In this regard, we propose a comprehensive conceptual model of how saline groundwater irrigation can conserve the artificial shelterbelt in the TDHS (Figure 4) [10]. Firstly, saline irrigation satisfies the plants' water needs despite salt accumulation at the soil surface, which minimally affects plants. To some extent, the planted trees can compensate for salt stress through adjustments in root morphology, such as the absence of feeder root growth within the salt accumulation layers. This adaption is one essential process for successful ecological engineering within arid environments. Secondly, the saline water irrigation may input nutrient components into the shelterbelt's soil, further synergizing with vegetative litter decomposition, root growth, and other biogeochemical cycles [23]. Like many other nutrient-constrained environments, litter decomposition within the TDHS is a critical biogeochemical process for C and nutrient cycling that has been primarily affected by the initial contents of C and K, lignin, and cellulose [24]. The accumulation of soil C and nutrients has generally increased throughout the sand-binding vegetation. Thirdly, because of the litter decomposition, microbial activity, and root exudates, the soil nutrient contents of C, N, and P have been significantly improved [25], all of which have contributed to soil development. Consequently, our artificial oasis may have developed into a more self-regulatory construct [16], characterized by the structured clay–organic-rich topsoil, restricted evaporation, and strengthened soil moisture regulation function. While water is the foundation of the oasis' survival, the soil is part of the synergy function for maintaining moisture. It has been shown that soil moisture and texture can affect weather, surface evaporation, and temperature extremes [26]. Furthermore, plant roots may develop tolerance to the local conditions, and can even reach the fringe of the groundwater. As evidenced, naturally distributed *Tamarix* cones, reliant on the relatively stable water sources, reflect adaptations to the different habitats [27]. Therefore, we suggest that shelterbelts and the aeolian sandy soil have developed synergistically over time, with more mature shelterbelts and biodiversity, better soil fertility, and reduced salinity. All of these factors, when considered collectively, have created a benign cycle.

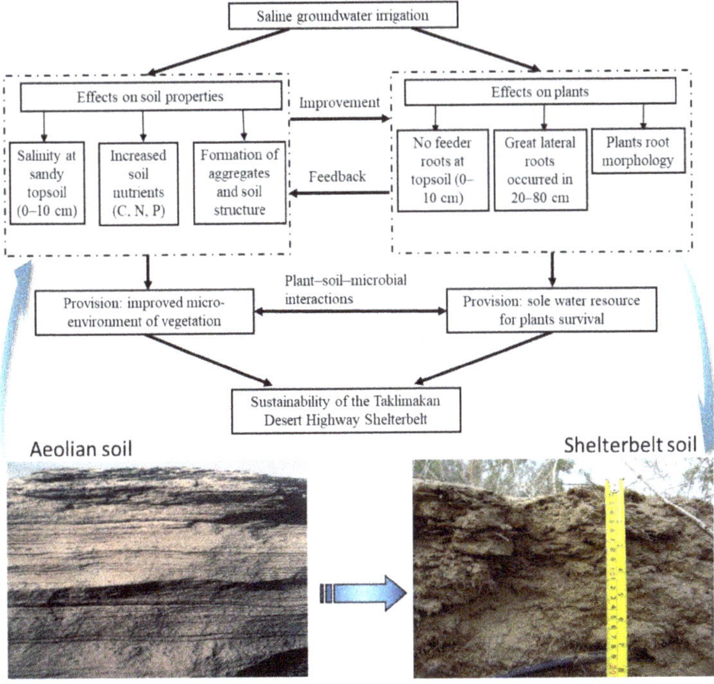

Figure 4. A conceptual model of a plant–soil–microbial system with saline groundwater irrigation. The figure is referenced and updated from [10].

4.2. *Implications from China's "Green Wall"*

Our studies have shown the importance of controlling and reclaiming areas overtaken by desert. Drylands have the lowest biological productivity, and are home to a significant fraction of global poverty [28]. Over 27% of China now suffers from desertification, making desertification a significant national threat [29]. The Three-North Shelter Forest Program is the most extensive afforestation program globally. It is a 2800-mile network of forest belts covering all of the major deserts and sandy lands within Northwest China. The project is designed to serve as a windbreak to stop sandstorms, halt the expansion of desertification, and restore the land to a productive and sustainable state. The local government has been mobilizing the masses to grow trees from the very beginning and, subsequently, more people have joined the campaign, digging holes for trees and fertilizing the land. As such, there needs to be greater engagement with the land users themselves, who can implement practices that abate land degradation and desertification [30]. Engagement means both education and outreach, highlighting the links between agriculture and ecology, and using innovative strategies to involve stakeholders in gathering and using their local knowledge, thus establishing a new paradigm for water- and climate-smart land management.

Identifying pathways to break the vicious cycle between land degradation and poverty in dryland SESs requires an in-depth understanding of the relationship between the degradation of ecosystem services and the deprivation of humans [5]. Programs must be evidence-based, prioritize cost-effective interventions, and adapt their priorities and approaches over time. Since its establishment, the TDHS has accomplished a series of research and development achievements through long-term monitoring, research, experiments, and demonstrations, and has focused on the key scientific and technological issues urgently needed for management of the fragile regional ecological environment. This study offers a unique laboratory to understand saline and dry environments, showing that they are more sensitive to our interventions than previously thought [29].

5. Sustainable Evaluation of the TDHS

Our project—an ecological engineering feat rivaling the construction of China's Great Wall—may be a successful artifact of human adaption to nature. Beset by water scarcity and growing deserts, China has been actively researching afforestation to create a robust ecosystem-based adaptation for decades. Various pilot projects have tried to mitigate desertification and even desalinate seawater. While long-distance water transport might seem impractical, the Taklimakan "experiment" shows concrete and large-scale evidence that the seed of that transformation is technically feasible by employing local saline groundwater. However, it is expected that the large-scale afforestation may have significantly impacted water resources, e.g., causing increased evapotranspiration and decreased runoff [30]. In concert with China's unprecedented economic growth, ecological restoration within China reduces soil water storage at one of the world's highest rates [31]. Field studies have reported that afforestation has lowered local groundwater levels by between 0.5 and 3.0 m, which may have reduced the survival rates of afforestation to 7–34%, limiting its effectiveness in desertification control [32]. When the results are considered at large scales, the water impacts may be significant, e.g., the large-scale mismatch of plant species' water requirements. For example, afforestation in the Chinese Loess Plateau is approaching sustainable water-resource-use limits and threatening local water and food security [31]. Consequently, China should take bold steps to implement sustainable development for large-scale green engineering.

Can we seek inspiration from ancient water engineering projects that did not use mechanical systems, such as the legendary Hanging Gardens of Babylon or the historic Roman water designs? Indigenous and local knowledge is slowly changing societal water-use perceptions. Close to the TDHS, Xinjiang Karez—an irrigation system of wells connected by an underground channel—is still used today, having been constructed back during the Han Dynasty about two thousand years ago [33]. The synergy of those ancient water delivery systems was planned and constructed without computer-aided design or mechanical systems, but these projects rival and perhaps even exceed the success of many similar designs based on 21st-century innovations [34]. We hope that the intelligent passive systems of the past can be resurrected.

Importantly, no significant long-term impacts of irrigation were found concerning the groundwater level and its salinity [35]. Within the TDHS, the local saline groundwater has been used with our specially designed double-branch-pipe drip-irrigation system, which ensures effective and efficient groundwater use [13] (Figure 2). The borehole wells have been placed at a distance of every 4 km for use on a 2-week irrigation schedule with 20–30 mm of groundwater. Hence, the maintenance cost is mainly used to irrigate and maintain the shelterbelt, at a cost of around USD 0.1 million km^{-1} a^{-1} (Table 1). Although the cost of maintenance is much more than the funding available based on a vehicle tax, this cost has been quickly reduced to >50% of the expense before the TDHS, mainly used to clean the shifting sand and construct the straw checkerboard barriers [36]. The TDHS is of the highest priority for oil/gas exploration and local community development, with a significant eco-environmental value that has not been well evaluated. It is expected to further decrease the cost with the progress of ecosystem services. At present, stable investment mechanisms for combating desertification need to be established, along with financial support policies for guiding the country in its fight against desertification [33]. It is important not to lose sight of these critical issues for regional development in the post-COVID-19 world, as resources and priorities may be shifted away from these crucial development arenas in desert regions.

Table 1. Sand fixation efficiency and cost of the three sand-fixing measures [36].

Control Measures	Effects of Sand Prevention	Input (USD km^{-1} a^{-1})	Ecological Benefits
Mechanical measures	More than 70% of the control measures were ineffective	~0.1 million	No
Biological measures	More than 90% of the artificial shelterbelts were effective	~0.2 million	Improve the eco-environment and increase biodiversity
Chemical measures	Ineffective	~0.15 million	Bring chemical materials to the desert and damage the eco-environment

Nonetheless, dryland sustainability programs need to ensure that the water requirements of species used in the large-scale restoration are compatible with local environmental water availability and quality [5]. If incorrectly managed in desert areas, the increasing aridity, enhanced warming, and rapidly growing population may exacerbate the risk of desertification [37]. Although our practices demonstrate that drip-irrigation is the optimal method for low-cost and highly efficient water usage, our approach needs a long-term assessment, considering that plant salt sensitivity is different for different species. For instance, *Calligonum* is not a halophyte genus, but a xeromorphic genus more susceptible to salt damage. The introduction of an alternative plant may be necessary, and reasonable irrigation methods associated with the growth of halophytes are suggested [38–40]. Only when careful choices are made concerning the irrigation water salinity and the plant species' tolerance for salinity can artificial shelterbelt construction promise sustainable development.

6. Technology Transfer

The TDHS represents a remarkable achievement of land-system sustainability with respect to scientific evidence, governance, and human endeavors [29]. Different regions/countries can learn from the TDHS project and, possibly, adopt similar methodologies to construct sustainable transportation routes or barriers against shifting dunes. With the launching of "the Belt and Road" program, the relevant techniques of our model have been the basis of technical training for transfer to similar habitats that have sustainable development needs for the economy, environment, and ecosystem health. We have currently shared our knowledge on desertification control with some African and Asian countries, e.g., Mauritania, Ethiopia, and Kenya, among others. For example, we proposed an optimization plan named "integrated construction system of 2 zones with 3 protective belts" in Mauritania's capital city Nouakchott. Moreover, international platforms, such as the "China–Africa Cooperation Forum", have established a cooperative mechanism for regularly exchanging experiences, techniques, and investments in desertification control. Some of these forums have been well recognized in a UN resolution as an essential platform for realizing the strategic objectives of the Convention on Combating Desertification.

Current large-scale evidence suggests that, despite enormous investments, China's integrated portfolio of sustainability programs has achieved overall success. Nations need to be prepared to take urgent, decisive, and robust action to ensure environmental sustainability. Before decisions on extensive water resource projects are made, it is necessary to summarize the local ecological, economical, and indigenous knowledge on nature's contributions to people, along with an appropriate decision-making framework of the dryland development paradigm for ecosystem services [41]. Moreover, research on the structure and function of dryland SESs requires sufficient attention. Delgado-Baquerizo et al. [42] found that, regardless of soil age, global climatic and land-use changes will have substantial long-term impacts on the structure and function of terrestrial ecosystems. In terms of both past and present perspectives, as exemplified by the TDHS program's effects across multiple sustainability indicators, our research may provide some keys to success from China's experience, discuss potential risks to large-scale sustainability interventions, and suggest future research priorities.

7. Concluding Remarks

The TDHS project offers the first proof of how human-aided establishment of vegetation on hyper-arid mobile dunes, supported solely by local saline groundwater, can be achieved. Some key points of complex adaptive systems are as follows: (1) Practically, local saline water irrigation within the TDHS shelterbelt has proven helpful for the healthy growth of adaptive plants, and also for the evolution of sandy soil. (2) Our conceptual model provides intrinsic driving mechanisms and structure–function interactions for a catastrophic regime shift of bare sand into vegetated soil. (3) If adequately managed, the TDHS will sustainably represent a successful model utilizing ecological engineering in a harsh environment, given the necessity of plant survival and a self-regulatory adaptation function. As a research agency, we continue to identify research areas needing critical evaluation, assess and verify program effectiveness, and provide scientific and technological support to implementation agencies for quality assurance of sustainability interventions against accepted standards. This program will be revisited and updated regularly to reflect research priorities and new requirements in drylands. We hope that the TDHS will be remembered not only as an engineering accomplishment, but also as a unique system of ecological achievements and a successful example of a large-scale green desert that other arid desert regions can follow.

Author Contributions: Conceptualization and methodology, Y.Z. and N.W.; formal analysis, Y.Z.; investigation, J.X. and Y.Z.; resources, Y.Z.; writing—original draft preparation, Y.Z.; writing—review and editing, J.X., Y.Z., N.W. and R.L.H.; project administration, Y.Z. All authors have read and agreed to the published version of the manuscript.

Funding: This research was supported by the National Natural Science Foundation of China (41977009); the National Talents Project (Y472241001); the cooperative program of the CAS key laboratory; the original innovation project of the basic frontier scientific research program, Chinese Academy of Sciences (ZDBS-LY-DQC031); and the Youth Innovation Promotion Association of the Chinese Academy of Sciences (2019430).

Acknowledgments: The authors would like to thank the Taklimakan Desert Research Station for the research outcomes and field practices involved in this paper. We also thank the anonymous reviewers and editor for their great help in improving the quality of the manuscript.

Conflicts of Interest: The authors declare no conflict of interest.

References

1. Xu, X.; Li, B.; Wang, X. Progress in study on irrigation practice with saline groundwater on sand lands of Taklimakan Desert hinterland. *Chin. Sci. Bull.* **2006**, *51* (Suppl. I), 161–166. [CrossRef]
2. Scanlon, B.R.; Bongers, F.; Lambin, E.F. Ecological controls on water-cycle response to climate variability in deserts. *Proc. Nat. Acad. Sci. USA* **2005**, *102*, 6033–6038. [CrossRef] [PubMed]
3. Eggenberger, S.; Gobet, E.; van Leeuwen, J.F.N. Millennial multi-proxy reconstruction of oasis dynamics in Jordan, by the Dead Sea. *Veget. Hist. Archaeobot.* **2018**, *27*, 649–664. [CrossRef]
4. United Nations. *Transforming Our World: The 2030 Agenda for Sustainable Development*; United Nations: New York, NY, USA, 2015.
5. Fu, B.; Smith, D.M.S.; Lambin, E.F. The Global-DEP conceptual framework—Research on dryland ecosystems to promote sustainability. *Cur. Opin. Env. Sus.* **2021**, *48*, 17–28. [CrossRef]
6. Li, Y.; Bao, W.K.; Bongers, F.; Chen, B.; Chen, G.K.; Guo, K.; Jiang, M.X.; Lai, J.S.; Lin, D.M.; Liu, C.J.; et al. Drivers of tree carbon storage in subtropical forests. *Sci. Total Environ.* **2019**, *654*, 684–693. [CrossRef]

7. Lei, J.Q.; Li, S.Y.; Jin, Z.Z.; Fan, J.L.; Wang, H.F.; Fan, D.D.; Zhou, H.W.; Gu, F.; Qiu, Y.Z.; Xu, B. Comprehensive eco-environmental effects of the shelter-forest ecological engineering along the Tarim Desert Highway. *Chin. Sci. Bull.* **2008**, *53* (Suppl. II), 190–202. [CrossRef]
8. Zhang, J.; Xu, X.; Lei, J.; Sun, S.; Fan, J.; Li, S.; Gu, F.; Qiu, Y.; Xu, B. The salt accumulation at the shifting aeolian sandy soil surface with high salinity groundwater drip irrigation in the hinterland of the Taklimakan Desert. *Chin. Sci. Bull.* **2009**, *53*, 63–70. [CrossRef]
9. Zhang, J.G.; Lei, J.Q.; Wang, Y.D.; Zhao, Y.; Xu, X.X. Survival and growth of three afforestation species under high saline drip irrigation in the Taklimakan Desert, China. *Ecosphere* **2016**, *7*, e01285. [CrossRef]
10. Li, C.J.; Lei, J.Q.; Zhao, Y.; Xu, X.W.; Li, S.Y. Effect of saline water irrigation on soil development and plant growth in the Taklimakan Desert Highway Shelterbelt. *Soil Till. Res.* **2014**, *146*, 99–107. [CrossRef]
11. Danin, A.; Bar-Or, Y.; Dor, I.; Yisraeli, T. The role of cyanobacteria in stabilization of sand dunes in Southern Israel. *Ecol. Medit.* **1989**, *XV*, 55–64. [CrossRef]
12. Liu, J.; Zhao, Y.; Zhang, J.; Hu, Q.; Xue, J. Effects of Irrigation Regimes on Soil Water Dynamics of Two Typical Woody Halophyte Species in Taklimakan Desert Highway Shelterbelt. *Water* **2022**, *14*, 1908. [CrossRef]
13. Zhang, J.; Xu, X.; Li, S.; Zhao, Y.; Zhang, A.; Zhang, T. Is the Taklimakan Desert Highway Shelterbelt sustainable to long-term drip irrigation with high saline groundwater? *PLoS ONE* **2016**, *11*, e0164106. [CrossRef] [PubMed]
14. Han, W.; Cao, L.; Yimit, H.; Xu, X.W.; Zhang, J.G. Optimization of the saline groundwater irrigation system along the Tarim Desert highway ecological shelterbelt project in China. *Ecol. Eng.* **2012**, *40*, 108–112. [CrossRef]
15. Qadir, M.; Steffens, D.; Yan, F.; Schubert, S. Proton release by N_2-fixing plant roots: A possible contribution to phytoremediation of calcareous sodic soils. *J. Plant Nutr. Soil Sci.* **2013**, *166*, 14–22. [CrossRef]
16. Liang, J.; Shi, W. Cotton/halophytes intercropping decreases salt accumulation and improves soil physicochemical properties and crop productivity in saline-alkali soils under mulched drip irrigation: A three-year field experiment. *Field Crop. Res.* **2020**, *262*, 108027. [CrossRef]
17. Liu, J.; Zhao, Y.; Sial, T.A.; Liu, H.; Wang, Y.; Zhang, J. Photosynthetic responses of two woody halophyte species to saline groundwater irrigation in the Taklimakan Desert. *Water* **2022**, *14*, 1385. [CrossRef]
18. Wang, Y.D.; Zhao, Y.; Li, S.Y.; Shen, F.Y.; Jia, M.M.; Zhang, J.G.; Xu, X.W.; Lei, J.Q. Soil aggregation formation in relation to planting time, water salinity, and species in the Taklimakan Desert Highway shelterbelt. *J. Soil Sed.* **2017**, *18*, 1466–1477. [CrossRef]
19. Young, I.M.; Crawford, J.W. Interactions and self-organization in the soil-microbe complex. *Science* **2004**, *304*, 1634–1637. [CrossRef]
20. Yi, Z.J.; Zhao, C.H. Desert "Soilization": An ecomechanical solution to desertification. *Engineering* **2016**, *2*, 270–273. [CrossRef]
21. Xi, Y.Q.; Zhao, Y.; Li, S.Y. Effects of Three Kinds of Soil Amendments on Shear Strength of Aeolian Soil. *Acta Pedol. Sin.* **2018**, *55*, 1401–1410.
22. Xue, J.; Gui, D.; Lei, J.; Sun, H.; Liu, Y. Oasification: An unable evasive process in fighting against desertification for the sustainable development of arid and semiarid regions of China. *Catena* **2019**, *179*, 197–209. [CrossRef]
23. Schlesinger, W.H.; Raikes, J.A.; Hartley, A.E.; Cross, A.F. On the spatial pattern of soil nutrients in desert ecosystems. *Ecology* **1996**, *77*, 364–374. [CrossRef]
24. Zhang, X.; Wang, Y.; Zhao, Y.; Xu, X.; Hill, R.; Lei, J. Litter decomposition and nutrient dynamics of three woody halophytes under surface and buried treatment in the Taklimakan Desert. *Arid. Land Man.* **2017**, *4*, 1–17.
25. Dong, Z.; Li, C.; Li, S.; Lei, J.; Zhao, Y.; Umut, H. Stoichiometric features of C, N, and P in soil and litter of Tamarix cones and their relationship with environmental factors in the Taklimakan Desert, China. *J. Soils Sed.* **2019**, *20*, 690–704. [CrossRef]
26. Fatichi, S.; Pappas, C. Constrained variability of modeled T:ET ratio across biomes. *Geophys. Res. Lett.* **2017**, *44*, 6795–6803. [CrossRef]
27. Dong, Z.; Li, S.; Zhao, Y.; Lei, J.; Wang, Y.; Li, C. Stable oxygen-hydrogen isotopes reveal water use strategies of Tamarix taklamakanensis in the Taklimakan Desert, China. *J. Arid Land.* **2020**, *12*, 115–129. [CrossRef]
28. Safriel, U.; Adeel, Z. Development paths of drylands: Thresholds and sustainability. *Sustain. Sci.* **2008**, *3*, 117–123. [CrossRef]
29. Cao, S.X.; Zhong, B.L.; Yue, H.; Zeng, H.S.; Zeng, J.H. Development and testing of a sustainable environmental restoration policy on eradicating the poverty trap in China's Changting County. *Proc. Natl. Acad. Sci. USA* **2019**, *106*, 10712–10716. [CrossRef]
30. Bryan, B.A.; Smith, D.M.S.; Lambin, E.F. China's response to a national land-system sustainability emergency. *Nature* **2018**, *559*, 193–204. [CrossRef]
31. Feng, X.; Venkatesan, K.; Balakrishnan, V. Revegetation in China's Loess Plateau is approaching sustainable water resource limits. *Nat. Clim. Change* **2016**, *6*, 1019–1022. [CrossRef]
32. Cao, S.X.; Lei, J.Q.; Xu, X.W. Excessive reliance on afforestation in China's arid and semi-arid regions: Lessons in ecological restoration. *Earth Sci. Rev.* **2011**, *104*, 240–245. [CrossRef]
33. Bhattacharji, P. Uighurs and China's Xinjiang Region. *Washington Post*, 1 August 2008; 1–2. Available online: http://www.washingtonpost.com/wp-dyn/content/article/2008/08/01/AR2008080100933.html (accessed on 1 June 2022).
34. Seckler, D.; Barker, R.; Amarasinghe, U. Water scarcity in the twenty-first century. *Int. J. Water Res. Dev.* **1999**, *15*, 29–42. [CrossRef]
35. Fan, J.L.; Xu, B.W.; Xu, X.W.; Li, S.Y. Technology Progress of Sand Industry Based on Desertification Control Engineering by Revegetation: Case Study on Cistanche Plantation. *Bull. Chin. Acad. Sci.* **2020**, *35*, 717–723.

36. Li, C.J.; Wang, Y.D.; Lei, J.Q.; Xu, X.W.; Wang, S.J.; Fan, J.L.; Li, S.Y. Damage by wind-blown sand and its control measures along the Taklimakan Desert Highway in China. *J. Arid Land.* **2021**, *13*, 98–106. [CrossRef]
37. Huang, J.; Li, Y.; Fu, C.; Chen, F.; Fu, Q.; Dai, A.; Rozema, J.; Flowers, T. Dryland climate change: Recent progress and challenges. *Rev. Geophys.* **2017**, *55*, 719–778. [CrossRef]
38. Ravindran, K.C.; Venkatesan, K.; Balakrishnan, V.; Chellappan, K.P.; Balasubramanian, T. Restoration of saline land by halophytes for Indian soils. *Soil Biol. Biochem.* **2007**, *39*, 2661–2664. [CrossRef]
39. Rozema, J.; Flowers, T. Crops for a salinized world. *Science* **2008**, *322*, 1478–1480. [CrossRef]
40. Wang, L.; Zhao, Z.Y.; Zhang, K.; Tian, C.Y. Reclamation and utilization of saline soils in arid northwestern China: A promising halophyte drip-irrigation system. *Environ. Sci. Technol.* **2013**, *47*, 5518–5519. [CrossRef]
41. Reynolds, J.F.; Smith, D.M.S.; Lambin, E.F.; Rozema, J.; Flowers, T. Global desertification: Building a science for dryland development. *Science* **2007**, *316*, 847–851. [CrossRef]
42. Delgado-Baquerizo, M.; Reich, P.B.; Bardgett, R.D. The influence of soil age on ecosystem structure and function across biomes. *Nat. Commun.* **2020**, *11*, 4721. [CrossRef]

MDPI
St. Alban-Anlage 66
4052 Basel
Switzerland
Tel. +41 61 683 77 34
Fax +41 61 302 89 18
www.mdpi.com

Water Editorial Office
E-mail: water@mdpi.com
www.mdpi.com/journal/water